RANCHING
HERITAGE
BOOKS

Families In Ranching Succession Toolkit

Building Your Legacy

Edited by Anish Quenim &
Claudio G. Muller

Foreword by Matt Allen

TEXAS TECH UNIVERSITY PRESS

This book is typeset in EB Garamond. The paper used in this book meets the minimum requirements of ANSI/NISO Z39.48-1992 (R1997). ♾

Designed by Hannah Gaskamp
Cover design by Hannah Gaskamp

Library of Congress Cataloging-in-Publication Data

Names: Quenim, Anish editor | Muller, Claudio G. editor Title: Families in Ranching Succession Toolkit (FIRST): Building Your Legacy / edited by Anish Quenim and Claudio G. Muller.
Description: Lubbock: Texas Tech University Press, 2026. | Series: Ranching Heritage Books | Includes bibliographical references and index. | Summary: "A practical guide to navigating family dynamics, estate planning, and legacy transfer in agricultural, and particularly ranching, enterprises"—Provided by publisher.
Identifiers: LCCN 2026020138 (print) | LCCN 2026020139 (ebook) |
ISBN 978-1-68283-329-2 paperback | ISBN 978-1-68283-330-8 ebook
Subjects: LCSH: Ranching—Management | Family-owned business enterprises—Management | Inheritance and succession | Farm management
Classification: LCC SF196.U6 F36 2026 (print) | LCC SF196.U6 (ebook)
LC record available at https://lccn.loc.gov/2026020138
LC ebook record available at https://lccn.loc.gov/2026020139

Texas Tech University Press
Box 41037
Lubbock, Texas 79409-1037 USA
800.832.4042
ttup@ttu.edu
www.ttupress.org

Contents

Illustrations

Tables

Foreword

Ranching families represent a unique "brand" (sorry, couldn't resist!) of family enterprise, and thus a unique set of succession-related decisions reflects universal challenges. The FIRST book covers the full range of related topics while providing highly useful concepts, frameworks, and tools for families navigating succession. The power in this book is that it specifically addresses ranching families, a topic that has been grossly underrepresented in literature on family-owned businesses.

A critical theme in this space is the inherent presence of emotion in decision-making, whether about succession, ownership, divestment, or other pertinent topics. The first chapter, "Understanding Evidence, Terminologies, and Business Family Dynamics," provides perspective and tools to recognize and mitigate this reality, be it centering succession-planning on data (versus tradition or individuals' agendas) or using clear governance structures and mediation strategies to respect emotional attachments while seeing beyond them.

Legacy, as covered in the "Communicating Legacies Through Goals" chapter, is a related, powerful force for families. My research and work with family enterprises illuminate that legacy can simultaneously be an anchor and an inspiration, potentially yielding entrenchment in past traditions or norms but also providing a flexible way forward as built on the family's values, as that chapter highlights through its focus on common challenges such as "Ruling from the Grave," which typically causes conflict and resentment. The chapter on "Budgeting, Financial Planning, Estate Planning, and Taxation," meanwhile, underscores the criticality of financial and legal planning—with future generations' needs in mind—as legs in the stool of succession, to provide for future generations, minimize tax burdens, and preserve family harmony.

No matter the level of foresight and planning, conflicts are inevitable, a central message of the "Mediating Associated and Unseen Conflicts" chapter. As the authors remind us, conflict has benefits, but the trick is recognizing conflict when it happens and taking proactive, practical steps to address it, whether through mediation or some alternative. “We will figure this out ourselves” is a common but risky mentality among families facing conflict, and the chapter’s frameworks and references represent useful means of digging into and resolving the issues at hand.

I found the chapter on “Shaping Milestones, Road Maps, and Family Meetings” particularly intriguing for its deep psychological approach to helping families shape flexible milestones to make and gauge succession progress. For example, anxiety, resistance, and decision fatigue are common reactions to these transitions and related processes, driven by a sense of loss of control and identity along with conscious and unconscious roles and dynamics. This can be particularly challenging territory for families to navigate, especially those in more traditional economic sectors like agriculture. Families would do well to heed the authors’ practical tips around active listening, validation, and de-escalation approaches, to “saddle up” more fully for the journey.

In summary, this FIRST book represents a unique resource for all generations of ranching families. The diversity of expertise among the authors, combined with the practical approach and engagement of real ranching families, ensures that this will be a trusted resource for ranching families now and into the future.

DR. MATT ALLEN, JOHN L. WARD CLINICAL PROFESSOR OF FAMILY ENTERPRISE
EXECUTIVE DIRECTOR, JOHN L. WARD CENTER FOR FAMILY ENTERPRISES
KELLOGG SCHOOL OF MANAGEMENT, NORTHWESTERN UNIVERSITY

Synopsis

F*amilies in Ranching Succession Toolkit (FIRST)* offers a comprehensive guide for ranching families navigating the complex process of intergenerational transfer. The book emphasizes that succession is more than just a business or financial transaction; it's a deeply emotional and cultural journey involving the transfer of legacy, knowledge, and identity. The toolkit provides frameworks, concepts, and practical tools to incoming, and outgoing, generations, and attempts to address the unique challenges faced by ranching families. Key themes include understanding business family dynamics, communicating legacies through clear goals, and navigating financial, estate, and taxation implications. It highlights the importance of evidence-based planning to mitigate bias and foster trust and offers strategies for mediating conflicts and developing clear milestones for the transition. The goal of *FIRST* is to help families initiate conversations, develop robust plans, and ultimately "keep the ranch in the family and the family in the ranch." The book aims to support ranching families in maintaining the "cultural continuity" of ranching for future generations.

A Word from the Editor

I didn't just grow up in a family business, I grew up in the heart of one, shaped by both its promise and its pain. My family owns a farm in India, a place that was once full of life, potential, and memories. As a youngster, I loved the ranch life. The land brought us together, connected us to nature, and reminded us of who we were. But over time, family conflicts and competing visions for the future unraveled that bond. That experience left a mark and a lesson. May I add that my learnings have helped heal most bonds with my family; hence my writing!

Now, living in the United States, I serve as the Director of the Alderson and Griffin Center for Family Business and teach Family Enterprise at the Rawls College of Business. Every day, I work with business families, students, and successors who are trying to do what my family once struggled to self-achieve: protect both the business and the relationships that make it meaningful.

Families in Ranching Succession Toolkit (FIRST) is a product of that journey. It is grounded in lived experience, my own and that of countless ranching families who face the emotional and practical challenges of succession. This book was not written from a distance. It is rooted in stories of conflict, reconciliation, experience, closure, and embracing the art of moving on together. After all, family is forever. It is for families who work the land together, who carry not just assets but ancestry, and who want to pass on something far deeper than profit, that "sense of legacy."

This toolkit offers guiding questions, reflection exercises, and strategies designed to help ranching families have hard conversations, make wise transitions, and walk the difficult path of succession with greater clarity and compassion. It does not promise easy fixes. But it does offer something I wish my family had access to back then: a structured, thoughtful way to talk, plan, and lead together. To the families reading this, here's a promise: "You'll never walk alone."

As you begin using this book, I also want to invite you to continue this journey with us. The Alderson and Griffin Center for Family Business and

Entrepreneurship will be offering ongoing support for the FIRST program with the National Ranching Heritage Center, an evolving platform for professional development, peer learning, and community engagement. Whether through workshops, conferences, or continuing education, we hope to walk with your family beyond these pages and provide a place for growth, dialogue, and success.

To conclude, I would like to thank all who helped make *FIRST* possible. Dr. Matt Allen. Our co-authors: Dr. Claudio Muller, Dr. Erica Irlbeck, Dr. Fanni Coward, Dr. Patrick Hughes, Dr. Kelly Lange, Dr. Mitzi Lauderdale, Dr. Jason Headrick, Mr. Gene Valentini, and Mrs. Amita Quenim. Our student assistants: Ms. Leigh Wells, Mr. Xan Argo, Mr. Barrett Moreland, and Mr. Rudy Benavides. Our associated university units: Mr. Jim Bret Campbell and the team at the National Ranching Heritage Center; the Ranching Heritage Association; Dr. Rod Williams and the Office of the Provost Outreach and Engagement; and Dr. Travis Snyder, Ms. Joanna Conrad, and all at Texas Tech University Press. And the Office of Dispute Resolution, Lubbock County, and icebreaQers, Goa-India, whose contributions were invaluable to the creation of this book, and all that follows.

Special thanks to my family: the Quenims, the Moussas, the Dhurandhars, the Jayakars, the Gudelmans, and the Goffs for helping me to understand the true meaning of "family." And to Mr. David Alderson and family, Mr. Mark Griffin and family, Mr. Brian McCoy and family, and all at the Alderson and Griffin Center for Family Business and Entrepreneurship, Dean Dr. Margaret Williams, Dr. Don Kluemper, and fellow faculty, staff, and my current and former students in the Rawls College of Business at Texas Tech University—thank you for our togetherness, vision, persistence, and unity. Here's to our own Red Raider legacy! #WreckEm

ANISH QUENIM, FAMILY ENTERPRISE FACULTY AND DIRECTOR
ALDERSON AND GRIFFIN CENTER FOR FAMILY BUSINESS AND ENTREPRENEURSHIP
RAWLS COLLEGE OF BUSINESS, TEXAS TECH UNIVERSITY

Introduction

"The people of cow country developed a unique set of values," says ranching ambassador and songwriter Red Steagall. "Those values are honesty and integrity, loyalty, work ethic, dedication to your family, conviction about your belief in God, and practicing respect and common decency for your fellow man every day you live."

Ranchers continue to deal with challenges from a wide spectrum of sources, but they meet those challenges with a deep commitment to those values. For many of them, those cow country values are handed down through generations, along with a love for the land and the grit and determination required to hang on to it.

One of the greatest challenges in ranching today is the generational succession of both managing the ranch operation and developing plans for the land itself to be passed to the next generation. Navigating the business, tax, and cash flow implications of a succession can be daunting within any business. The process presents a wide array of questions and the need for quality expertise and counsel. The succession process can be exponentially more challenging when you mix in long-established dynamics of a family-owned and family-operated ranch.

Unfortunately, without a way to begin the conversation and a process for planning succession, many family ranches fall victim to circumstances. At the death of a patriarch or matriarch, the only recourse is to sell the ranch. Without a plan, the family might lose the land and, often, fractures relationships between family members that may never be reconciled.

The Families in Ranching Succession Toolkit (FIRST) program enables families to begin the conversation about estate, tax, and management succession. FIRST offers ranching families with a guide to asking the right questions, while acknowledging family dynamics, and provides resources on how to get started. FIRST will open the door to meaningful, impactful conversations designed to "keep the ranch in the family and the family in the ranch," as shared by our friends at the R. A. Brown Ranch in Throckmorton, Texas.

At the National Ranching Heritage Center, we tell the story of ranching over the past 200 years in the American West. We are excited to partner with the Alderson and Griffin Center for Family Business and Entrepreneurship at Texas Tech to share tools and resources with ranching families.

Our goal is to continue to tell the story of ranching families for another 200 years . . . and beyond.

JIM BRET CAMPBELL, EXECUTIVE DIRECTOR
NATIONAL RANCHING HERITAGE CENTER,
TEXAS TECH UNIVERSITY

Families In Ranching Succession Toolkit

CHAPTER 1

Understanding Evidence, Terminologies, and Business Family Dynamics

Claudio G. Muller

INTRODUCTION

The succession process in family ranching is far more than a transfer of assets—it is a transfer of legacy, knowledge, identity, and the values that define generations of stewardship and resilience. For many ranching families, the land is not simply a productive resource but also a living landscape imbued with stories, memories, and responsibilities. The cultural heritage embedded in family ranches, though often overlooked, is a vital aspect of sustainable rural development. This chapter begins with recognizing this cultural depth and continues with a practical toolkit for navigating the complex dynamics of business, kinship, and intergenerational continuity.

The heart of successful succession intertwines tangible planning mechanisms, such as financial documentation, leadership development, and governance structures, with intangible elements like experiential learning, folk traditions, and local ecological knowledge. These dimensions cannot be separated: To plan for succession meaningfully, one must embrace the family enterprise's business strategy and cultural narrative. As noted in Kirner's (2015) ethnographic study of Eastern Sierra ranchers, continuity depends not only on who will take over the ranch but also on how that successor has been immersed in the practice, values, and worldview of ranching itself.

Research shows that succession failure often stems not from external shocks but rather internal fragmentation, gaps in communication, misaligned expectations, lack of preparation for leadership, and a failure to honor the cultural and emotional attachments that underpin a family's connection to its land. Only a minority of family ranches in the United States successfully transition into the third generation. In ranching communities—where fewer than 2 percent of Americans maintain active agricultural livelihoods—preserving cultural continuity has implications for family wealth and identity and the resilience of entire rural ecosystems (USDA 2017).

This chapter explores succession planning from an evidence-based and culturally aware perspective. It provides families with frameworks for self-assessment, goal alignment, role clarity, and value articulation. It also challenges the reader to think beyond efficiency and profitability toward meaning, belonging, and legacy questions. Drawing on research from management studies and cultural anthropology, we argue that viewing the ranch as a heritage institution as well as a business enhances strategic thinking, builds intergenerational trust, and strengthens the capacity of families to adapt and thrive amid uncertainty.

By foregrounding evidence and tradition, this chapter encourages ranching families to adopt transition plans that are viable and meaningful. We aim to support ranching families in maintaining what Kirner (2015) calls the "cultural continuity" of ranching, an inheritance not just of land but of wisdom, values, and community stewardship.

THE IMPORTANCE OF EVIDENCE IN BUSINESS STRATEGY

In the rapidly evolving landscape of modern business, decisions grounded in well-researched evidence have become a cornerstone of strategic success. Evidence-based strategies provide businesses with a foundation to address uncertainties, leverage opportunities, and mitigate risks. This principle holds especially true for family businesses, where the interplay of familial relationships and business objectives creates a unique dynamic (Nordqvist and Melin 2010).

Evidence in business strategy refers to verifiable data, research findings, and insights derived from systematic analyses. According to Picone et al. (2021), these measures can include market trends, financial reports, customer feedback, and internal performance metrics. For family businesses, evidence also encompasses studies on succession planning, family dynamics, and governance structures. Such evidence ensures that decisions are based not merely on intuition or tradition but on objective factors that enhance the likelihood of success.

In ranch succession, where family expectations, emotional ties, and cultural identities are deeply entwined with land and labor, evidence is often perceived as clinical or impersonal. However, evidence-based planning can be a powerful tool to foster trust, clarify roles, and preserve family cohesion. Rather than replacing tradition, it grounds decisions in reality, making transitions more sustainable and just.

WHY EVIDENCE MATTERS

In family businesses, personal relationships and emotions often influence decisions. Evidence-based approaches help minimize bias by focusing on facts and data, ensuring that strategic choices serve the best interests of the business rather than individual preferences.

Evidence provides clarity and context, enabling decision-makers to predict outcomes more accurately. For instance, data on market demand can guide product diversification, while performance analytics can highlight areas requiring operational improvement.

Presenting clear, objective insights and evidence helps align family and business goals. For example, financial analyses might reveal the need for reinvestment in the business, which can be balanced against family members' financial aspirations.

Evidence-based decisions inspire trust among stakeholders, including employees, clients, and family members. Transparent processes backed by reliable data demonstrate professionalism and a commitment to informed decision-making.

Business landscapes are dynamic, influenced by technological advances, regulatory changes, and evolving consumer behaviors. Evidence allows family businesses to remain agile, adapting strategies to meet emerging challenges and opportunities.

WHY EVIDENCE MATTERS FOR SUCCESSION

- **Supporting Intergenerational Dialogue**. Evidence provides a neutral ground for difficult conversations. For example, presenting side-by-side ranch operating budgets can depersonalize disagreements about resource allocation between retiring and entering generations.
- **Facilitating Role Transitions**. Tracking the hours worked, decisions made, and responsibilities held by each family member allows

families to clarify who is ready for ownership roles and how to compensate them fairly.

- **Compensating the Returning Heir**. A family documented five years of labor and below-market wages received by the successor. This record was used to justify a larger equity share in the ranch LLC, preventing future disputes over "sweat equity."
- **Balancing On- and Off-Farm Heirs**. In a Montana ranching family, the successor worked on the ranch full-time while siblings pursued urban careers. The family used a business valuation and future cash flow projections to compensate the on-ranch heir with operational control while equalizing inheritance with insurance proceeds for off-ranch siblings.

Emotion and legacy are essential in ranching succession, but so is clarity. Evidence empowers families to make informed decisions without erasing the meaning behind them. By integrating complex data with shared stories and values, evidence-based strategy becomes not a replacement for tradition but a reinforcement of it, ensuring continuity and adaptability for future generations.

GUIDING SUCCESSION PLANNING

Succession planning is one of the most critical and complex challenges family businesses face, particularly those rooted in agriculture and ranching. It is more than a legal or financial transaction; it is a social and emotional process that unfolds across years and generations. Effective succession ensures the continuity of leadership, preserves the family's legacy, and secures the long-term sustainability of both the business and its cultural identity.

Achieving this meaningful and lasting transition requires both intention and structure. While no two families are alike, some consistent practices have emerged as effective across various ranching contexts. These practices provide a foundation for families to build customized succession plans that reflect their unique values, goals, and circumstances. Below, we outline key components of this process—each informed by research, lived experience, and cultural awareness that can help transform succession from a difficult obligation into a shared opportunity.

1. **Identifying Potential Successors**. Objective evaluation of potential successors is crucial. Families should use a combination of:

- Performance metrics, including years of experience, initiative, and problem-solving demonstrated on the ranch.
- Leadership potential, gauged through peer feedback, ability to manage people, and emotional intelligence.
- Commitment to values. Does the successor respect the ranch's traditions and vision for the future? For instance, a Wyoming family assessed all adult children using a leadership matrix that included ecological knowledge, business literacy, and collaborative decision-making skills.

2. **Leadership Development Programs**. A planned succession needs time. Structured development programs might include:

   - Mentorship from outgoing leaders and shadowing during critical seasons like calving or budgeting.
   - Education, in business management, regenerative practices, or marketing.
   - Rotations: exposure to each department of the ranch operation. One Colorado ranch created a three-year track where the next-generation leader managed one key domain annually starting with herd health, budgeting, and land stewardship.

3. **Addressing Family Dynamics**. Succession surfaces deep emotional dynamics. Facilitation and structured dialogue can:

   - Uncover unspoken expectations.
   - Identify sources of tension, e.g., rivalry, fairness, legacy.
   - Align personal goals with business needs. A Utah family hosted a professionally facilitated retreat where members shared their vision and fears about succession. The result: a shared purpose statement that now anchors their transition plan.

4. **Documenting the Succession Plan**. Transparency is key. A well-documented plan should include:

   - Timeline of role changes.
   - Governance structure (e.g., who makes decisions, how

conflicts are managed).

- Legal frameworks such as LLCs or family trusts.
- Communication protocols for regular updates and course corrections. A Nevada ranch drafted a "Family Charter" to supplement their legal documents, outlining roles, values, decision-making processes, and criteria for entry or exit.

Succession planning is most effective when it aligns operational strategy with family identity and values. To do so, families must first recognize themselves as business families—not just relatives working together but a unit whose identity is shaped by the intersection of kinship, enterprise, and cultural heritage.

This concept is further explored in Sidebar 3: The Importance of Recognizing Your Family Business Identity, which helps families understand how clarifying their identity can guide strategic decisions, resolve conflict, and build a sense of shared purpose (Heck et al. 2006). By explicitly acknowledging family members' dual roles—as business leaders and cultural stewards—ranching families can design succession plans that reflect who they are, not just what they do.

SIDEBAR 1: TERMINOLOGIES AND CONTEXT IN FAMILY DYNAMICS

Succession in ranching families is both a practical and a deeply personal journey. It involves legal, financial, and operational shifts and emotional and cultural transitions across generations. However, too often, families stumble not because they disagree on goals but because they misunderstand each other's language.

This section introduces key terms commonly used in family business and succession planning, which have been adapted specifically for the context of ranching. By learning and using a shared vocabulary, families can clarify intentions, avoid assumptions, and reduce miscommunication. This foundation supports more respectful conversations, transparent decision-making, and smoother transitions of leadership and ownership.

OWNERSHIP

These terms describe how ranch assets and decision rights are held and passed from generation to generation.

1. **Succession Planning**: A structured process of preparing the next generation to assume ownership and leadership, ensuring business continuity and preservation of legacy.
2. **Estate Planning**: Legal preparation for asset transfer after death, including wills, trusts, and inheritance strategies to minimize conflict and taxes.
3. **Buy-Sell Agreement**: A legal contract determining how ownership shares may be transferred in events such as death, disability, or retirement.
4. **Fairness vs. Equality**: A distinction in succession discussions where "fairness" considers contributions and needs, while "equality" suggests an equal division of assets.
5. **Transition Timeline**: A clear schedule outlining when leadership, management, and ownership responsibilities will change.
6. **Family Constitution**: A written agreement outlining the family's vision, values, rules of conduct, governance structures, and expectations for ownership, employment, and succession. It serves as a "charter" for guiding behavior and decision-making.

FAMILY

These concepts reflect the emotional and relational aspects that influence decision-making in ranch succession.

7. **Active Heir**: A family member engaged in day-to-day ranch operations, usually considered for leadership succession.
8. **Non-Active Heir**: A family member who does not work in the business but retains financial or emotional ties to the ranch.
9. **Legacy Preservation**: The intentional effort to sustain the ranch's historical, cultural, and familial meaning for future generations.
10. **In-Law Inclusion**: The extent to which spouses or partners of blood relatives are involved in governance, succession, and ownership decisions.
11. **Family Council**: A family-only advisory group that discusses strategic matters, resolves conflicts, and ensures the family voice is integrated into decision-making.
12. **Conflict Resolution Mechanisms**: Agreed-upon strategies to resolve disputes constructively, such as facilitated dialogues or

third-party mediation.

13. **Family Identity**: The shared values, traditions, and purpose defining the family's relationship with the ranch. Recognizing this identity can strengthen cohesion and guide strategic decisions.

BUSINESS

These terms relate to the ranch as an economic enterprise and its long-term strategic success.

14. **Generational Transition**: The planned and staged handoff of operational responsibilities, decision-making authority, and management roles to the next generation.
15. **Family Governance**: The structures (like charters or councils) and rules that regulate how family members interact with the business and each other.
16. **Sweat Equity**: Time and labor invested by family members, especially before they hold formal ownership. Often considered in succession and compensation decisions.
17. **Vision Statement**: A written declaration of long-term goals and aspirations for the ranch that aligns family purpose with business direction.
18. **Cultural Stewardship**: The commitment to maintaining ecological knowledge, historical land practices, and local traditions to manage the ranch.
19. **Ownership Roles Matrix**: A planning tool that maps current and future roles within the family, showing who is responsible for what and when transitions occur.

When every member of a ranching family understands and uses the same terms—such as "family identity," "active heir," or "buy-sell agreement"—they build a shared framework for navigating succession with clarity and mutual respect. This common language does not eliminate disagreement but makes it easier to address differences without confusion or resentment. In short, when ranch families speak the same language, they can plan the same future—together.

BUSINESS FAMILY DYNAMICS

Family dynamics significantly influence the performance and sustainability of ranching businesses. The intricate interplay of relationships, values, and individual aspirations can drive success or create conflict. Understanding and managing these dynamics is essential for long-term business viability.

Shared values often form the foundation of family businesses. These values influence decision-making, customer relations, and internal culture. While they provide a strong sense of identity and purpose, differing interpretations of these values among family members can lead to friction. Establishing clear, collective principles helps unify the family's vision (Muller 2024).

In family businesses, individuals often play multiple roles—as family members, employees, or shareholders. This overlap can blur boundaries, leading to misunderstandings or favoritism. Clear governance structures and role definitions are critical for maintaining professionalism and fairness.

Conflicts in family businesses often arise from sibling rivalries, generational differences, or differing ambitions. Open communication channels, facilitated discussions, and third-party mediators can help resolve disputes and foster collaboration. The dynamics of leadership transitions often highlight the underlying family tensions. Transparent processes, supported by evidence-based evaluations, ensure that leadership decisions prioritize the business's needs while respecting family relationships.

A collaborative and mutually respectful culture strengthen the family and the business. Family meetings, shared goal setting, and inclusive decision-making build trust and alignment. Family businesses face unique external challenges, including market competition, regulatory changes, and societal expectations. Balancing these pressures with internal family dynamics requires strategic adaptability and resilience.

By addressing these dynamics thoughtfully, family businesses can leverage their unique strengths, such as trust, long-term orientation, and a shared commitment to success—to achieve enduring growth and harmony.

SIDEBAR 2: PRACTICAL CASE: THE WALKER FAMILY RANCH[1]

The Walker Family Ranch, located in Northwestern Oklahoma, had been stewarded by the Walker family for more than eighty years. As patriarch George Walker began preparing for retirement, he hoped to pass the ranch to his two children, Emily and James.

Emily managed the ranch's daily operations for more than a decade, focusing on soil health, herd management, and relationships with neighboring

1. This case has been anonymized to protect the family's identity; names and other personally identifiable information have been removed or altered.

ranchers. James, the younger sibling, had recently returned after working in agricultural finance in Texas. He brought new ideas about expanding into agritourism and investing in online direct-to-consumer beef sales.

Although both siblings were committed to the ranch's success, deep tensions emerged.

- **Strategic differences**. Emily prioritized tradition, sustainability, and maintaining the ranch's reputation in the local community. James wanted to modernize quickly, viewing new revenue streams as essential to keeping the ranch competitive.
- **Resentment and recognition**. Emily believed her hard labor and years devoted to the ranch merited her greater control or ownership. James, however, felt his business acumen was undervalued, especially since he had declined a lucrative job to return home.
- **Lack of defined roles**. George had always made decisions unilaterally, and no formal governance systems were in place. This left Emily and James unsure of their authority, which led to stalled projects and growing interpersonal strain.

The Walkers engaged a family business consultant who helped them implement a governance plan. They created a decision-making framework, held structured family meetings, and developed a vision statement reflecting preservation and innovation. Emily was named operations director, while James took charge of financial planning and business development.

To ensure future clarity, the family also drafted a Family Constitution that included role descriptions, succession plans, and protocols for conflict resolution.

The Walkers have leveraged their strengths to stabilize and grow the ranch. By turning tension into trust and establishing boundaries and shared purpose, they have built a more resilient business and a stronger family.

RANCHING SUCCESSION: KEEPING IT IN THE FAMILY

Ranching businesses, often passed down through generations, epitomize the integration of family and business. The succession process in ranching families is particularly intricate due to the emotional ties to land, the unique operational challenges of agriculture, and the interplay of tradition and modernity (Dallin et al. 2021; Silva and Anjos 2022).

For ranching families, the land is more than a business asset; it symbolizes heritage and identity. This deep emotional connection can complicate succession discussions, as family members may have differing visions for the land's future. Different generations often hold distinct views on ranch management (Purc-Stephenson et al. 2025). Older generations may prioritize preserving traditional practices, while younger members might advocate for modernization and innovation. Balancing these perspectives requires open dialogue and shared planning.

Ranching operations demand a clear division of labor, often involving multiple family members. Succession planning must address how roles will transition over time and ensure that successors are adequately prepared to handle operational and managerial responsibilities (Thorne 2021). On the other hand, the capital-intensive nature of ranching makes financial planning a critical aspect of succession. Questions of ownership, inheritance, and equitable distribution among heirs must be addressed transparently to prevent future conflicts.

Another key consideration is the role of entrepreneurial intention and the active pursuit of innovation within younger generations. Hauck et al. (2016) highlight that generational successors are more likely to drive sustainable success when empowered to act entrepreneurially rather than being expected to maintain the status quo. In many ranching families, the assumption that "the next generation will just take over" can undermine succession. Without structured preparation, shared leadership, and support for the successor's vision, younger family members may feel constrained by tradition, leading to disengagement or exit from the enterprise (Muller and Sandoval-Arzaga 2021; Gomez-Mejia et al. 2020).

The integration of entrepreneurship into ranch succession also aligns with the growing need for diversification. Families that support new ventures—such as ecotourism, direct-to-consumer meat sales, or conservation leasing—are often more resilient to environmental and market volatility. These entrepreneurial activities strengthen the business and give successors a sense of ownership and innovation, creating deeper engagement in the transition process.

Family commitment is also a critical determinant of succession success. Hauck et al. (2016) note that strong affective commitment among family members is positively associated with succession preparedness and long-term planning. Investing in relationships—through shared experiences, trust-building, and collaborative visioning—is just as important as legal or financial planning.

Successful ranching succession requires balancing the preservation of the family's legacy with the need to adapt to changing market and environmental conditions. This might involve diversifying income streams, adopting sustainable practices, or exploring new technologies (Wittman and Radakovich 2009). The most enduring transitions frame succession as a shared journey—not merely a handoff but a transformation shaped by all the participating generations.

SIDEBAR 3: THE IMPORTANCE OF RECOGNIZING YOUR FAMILY BUSINESS IDENTITY

A family business is typically an enterprise involving two or more family members, and the majority ownership or control lies within the family. These businesses often integrate the unique dynamics of family relationships with entrepreneurial decision-making. Family enterprises range from small, locally owned establishments to large multinational corporations. What sets them apart is blending family values, traditions, and goals with business operations. Recognizing your business as a family enterprise is not merely about classification; it is about understanding the interconnectedness of your family's influence and the opportunities and challenges this brings.

Acknowledging your identity as a family business is crucial for leveraging inherent strengths and addressing potential pitfalls. Family businesses are often characterized by long-term thinking, a deep commitment to legacy, and a strong sense of trust and loyalty among stakeholders. These attributes can give family enterprises a competitive advantage in cultivating customer relationships, building employee loyalty, and creating sustainable practices. However, failing to recognize the family's role in the business can lead to unaddressed conflicts, unclear succession plans, and missed opportunities to harness family cohesion for strategic growth.

Furthermore, identifying as a family business can open the door to resources, support networks, and specialized guidance designed for such enterprises. It allows entrepreneurs to align their business strategy with their family's shared vision and values, creating a more unified approach to decision-making. Additionally, it fosters transparency within the organization, making navigating sensitive issues like succession planning, governance structures, and conflict resolution easier. By embracing their identity as family entrepreneurs, business owners can build resilience and longevity for the business and the family's legacy.

a. Family Business Role Identification and Future Planning Matrix
   Purpose: To map the roles and responsibilities of family

members, owners, and employees to better understand a family business's current dynamics and future aspirations.

b. Rules of Entry and Exit Assessment Matrix
Purpose: To ensure clarity and harmony in family businesses, establishing clear rules of entry and exit for family members, owners, and employees is essential. This tool provides a structured matrix to assess and define these rules, supporting smoother transitions, reduced conflicts, and sustained business success.

c. Family and Business Goals and Performances Assessment Matrix
Purpose: Family businesses often juggle dual goals: achieving business success and maintaining family harmony. This assessment matrix, inspired by Sharma (2004) and Ward (1987), helps family businesses evaluate the importance and performance of their goals across family and business dimensions. Families can align efforts for sustainable growth and unity by identifying strengths and gaps.

Table 1.1. Family Business Role Identification and Future Planning Matrix

Role Category	**Current Role(s)**	**Challenges / Concerns**	**Future Role(s)**	**Development Actions Needed**
Family Members	Identify current involvement of family members in business (e.g., decision-making, advisory roles).	Highlight challenges such as conflicts, lack of involvement, or unclear boundaries.	Define aspirations for the next generation or changing roles (e.g., succession planning, mentorship roles).	Specify actions like training, mentoring, or conflict resolution.
Owners	Identify who owns the business (e.g., sole proprietorship, sibling partnership).	Note ownership-related challenges such as unclear succession plans or disputes.	Define future ownership structure (e.g., transition to next generation, co-ownership).	Define steps like succession planning, shareholder agreements, or governance improvements.

Role Category	Current Role(s)	Challenges / Concerns	Future Role(s)	Development Actions Needed
Employees	Map current roles of family and non-family employees (e.g., operational, strategic).	List challenges like favoritism, lack of skills, or unclear career paths.	Plan for future roles of employees (e.g., leadership transitions, external hires).	Define actions such as professional development, formal HR policies, or leadership training.
Family-Owners	Identify family members who are both owners and actively involved in the business.	Highlight conflicts, overburdened roles, or unclear separation between family and ownership.	Define desired future roles (e.g., board advisory, active leadership).	Actions may include succession planning, setting boundaries, or leadership development.
Family-Employees	Highlight family members working in the business but not in ownership roles.	Note concerns about career growth or fairness.	Define future roles (e.g., equity participation, leadership development).	Specify steps like education, external experience, or mentorship programs.
Owner-Employees	Identify non-family owners who are also employees.	List challenges such as alignment with family values or potential conflicts.	Define future roles (e.g., transition to advisory positions).	Actions include governance updates or improved communication strategies.

Table 1.2. Rules of Entry and Exit Assessment Matrix

Criteria	Current Rules	Challenges Identified	Proposed Rules	Actions Needed
Ownership Eligibility	Outline existing criteria for family ownership (e.g., inheritance, buy-in, or gifting shares).	Highlight gaps, such as clarity on minority vs. majority ownership rights.	Define rules for ownership transfer (e.g., timeline, payment terms, inclusion of non-blood relatives).	Develop shareholder agreements, communication plans, and legal structures for ownership transfers.
Employment Eligibility	Specify criteria for family members to work in the business (e.g., education, prior experience).	Note issues like favoritism, lack of qualifications, or unclear expectations.	Set minimum qualifications (e.g., degrees, work experience outside the business).	Create HR policies, develop mentorship programs, or organize external experience opportunities.
Performance Standards	Describe how family employees are evaluated (e.g., formal reviews, informal feedback).	Identify challenges like perceived unfairness or lack of accountability.	Establish performance evaluation frameworks with clear metrics and unbiased reviewers.	Train evaluators, set transparent KPIs, and ensure regular performance reviews.
Exit Rules for Owners	Specify conditions for selling or transferring ownership (e.g., approval processes, buy-back rights).	Note disputes over share valuation or unclear exit options for disinterested owners.	Define fair valuation processes and conditions for selling shares within or outside the family.	Draft buy-sell agreements, create liquidity reserves, and conduct periodic evaluations.
Exit Rules for Employees	Outline how family members leave employment (e.g., resignation terms, severance policies).	Highlight issues like unclear post-exit roles or family tensions.	Define exit protocols, post-exit restrictions (e.g., non-compete clauses), and transitional responsibilities.	Provide career counseling, retirement planning, and conflict resolution services for smooth transitions.

Table 1.3. Family and Business Goals and Performances Assessment Matrix

Dimension	Goals	Importance (Low / Med / High)	Current Performance (Low / Med / High)	Strengths Identified	Challenges Identified	Action Plan
Business	Growth in sales					
	Growth in market share					
	Growth in profitability					
	Return on equity					
	Employee turnover					
	Expansion into new markets					
Family	Family harmony					
	Reputation in the community					
	Development of the next generation					
	Growth in human capital					
	Wealth growth					

CONCLUSION AND LESSONS LEARNED

Ranching succession is more than an operational or financial transition—it is an intergenerational endeavor that calls for careful stewardship of both tangible assets and intangible values. This chapter emphasizes that successful succession planning in ranching families is anchored in three pillars: using evidence to inform decisions, developing a shared vocabulary to reduce miscommunication, and managing business family dynamics with structure and empathy.

By recognizing the ranch as a business and a legacy, families can engage in strategic planning that reflects their unique identity and aspirations. When combined with open dialogue and inclusive governance, evidence-based planning allows for more equitable outcomes and stronger generational bonds. Once seen as a potential source of tension, family dynamics can instead drive innovation, cohesion, and resilience.

1. Succession Is a Cultural and Strategic Act. It involves not just the transfer of land or leadership but the continuation of a way of life rooted in heritage, responsibility, and stewardship.
2. Shared Language Reduces Conflict. A common understanding of key terms (e.g., fairness vs. equality, active heir, family constitution) enables more productive conversations and reduces ambiguity.
3. Governance Structures Empower Families. Clarifying roles, responsibilities, and decision-making processes minimizes confusion and supports smoother transitions.
4. Evidence Builds Trust and Legitimacy. Data-driven approaches to valuation, performance, and role-readiness ground decisions in fairness and transparency, avoiding emotionally driven disputes.
5. Family Identity Shapes Strategy. Recognizing the family as a unique social and entrepreneurial unit helps align strategic planning with long-term family goals and values.
6. Entrepreneurial Intention Enhances Continuity. Supporting the next generation's entrepreneurial ideas (e.g., agritourism and conservation projects) increases engagement and keeps the business adaptable and future focused.

NEXT STEPS

To turn insights into action, ranching families are encouraged to:

- Initiate conversations. Begin open discussions about values, expectations, and future visions among generations.
- Conduct a Family Business Assessment. Use tools like the Family Business Role Matrix and Rules of Entry and Exit Matrix to map current roles, challenges, and aspirations.
- Develop a Family Constitution. Articulate shared values, governance rules, and long-term vision in a collaborative living document.
- Engage outside facilitation. Bring in advisors or mediators to support sensitive conversations and planning processes.
- Schedule regular succession check-ins. Set yearly family meetings to revisit succession plans, adapt to new conditions, and reaffirm shared goals.
- Encourage youth development. Provide leadership opportunities, mentorship, and support for next-generation members to grow into meaningful roles.

Ranching families can safeguard their heritage by taking these steps while positioning their operations and relationships for continued success across generations.

CHAPTER 2

Communicating Legacies Through Goals

Erica Irlbeck, Fanni Coward, and Patrick Hughes

CHAPTER SUMMARY

This chapter emphasizes effective communication and goal setting in legacy planning. It explores the emotional impact of sudden loss and the importance of establishing clear intentions for the future. The chapter also highlights the benefits of effective legacy planning, such as empowering future generations and strengthening family relationships, while addressing pitfalls like posthumous control and conflicts over inheritance. Practical tools and strategies for setting and communicating goals are provided, including tips for effective planning. Real-life stories illustrate successful and challenging legacy planning scenarios, offering valuable insights and lessons.

LEARNING OBJECTIVES

After reading this chapter, you should be able to:

- **Understand the Importance of Goal Setting**. Explain why setting clear goals is important in legacy planning.
- **Identify Legacy Giving Values**. Recognize the values that affect legacy planning decisions and communication.
- **Identify Legacy Planning Communication Strategies**. Outlining goals and values may contribute to successful legacy planning communication strategies.
- **Analyze Legacy Planning Scenarios**. Evaluate legacy planning examples of successful and unsuccessful legacy planning.

INTRODUCTION: IMPORTANCE OF GOAL SETTING IN LEGACY PLANNING

If you have experienced or come across accounts of individuals who have faced the heart-wrenching loss of a loved one, you understand that this is an intensely emotional and often bewildering time for family members. This grief can be overwhelming, leaving those left behind grappling with not only sorrow but also practicalities of managing a loved one's estate. In this context, goal setting in legacy planning is not only beneficial but essential. By clearly defining your intentions and wishes regarding your assets and responsibilities, you provide invaluable guidance to your family during a tumultuous period. When you take time to articulate your goals, be it in terms of financial resources, personal possessions, or charitable aspirations, you simplify the process for your loved ones and reduce stress for them. Ultimately, goal setting in legacy planning is a profound gift to your family, nurturing peace of mind and fostering a sense of unity as they carry forward your legacy.

LEGACY PLANNING IS EMOTIONAL, ESPECIALLY WHEN THE LOSS IS SUDDEN

Fanni's Story

Life is not certain, but death is inevitable. On the fateful date of February 4, 2023, I was struck by the chilling truth of this lesson. It started out as a beautiful day. I remember the sun poured golden light into the dining room area as we gathered to celebrate one of my "I-do" children's birthdays. We sat around the table and laughter spilled from every corner, weaving through the air like a playful breeze, while the familiar aroma of steaming ramen—the birthday boy's favorite meal—filled the room. Our hearts also brimmed with anticipation, and talk of our soon-to-be-born granddaughter danced among us like spring butterflies. Each mention of her name ignited and illuminated my husband's face with joy. I love seeing him with that smile and hearing his laughter, rich and full, a tangible warmth that wrapped around me like a blanket. I remember feeling the very essence of happiness swirling in the air. Then a sudden stillness enveloped us; my husband leaned his head upon my shoulder and the laughter faded. Little did I know that beautiful moment would slip away like sand through my fingers, and that I would be powerless to stop it. My husband was only 61, and in an instant the world I knew changed.

Since that day, the pain, confusion, and brokenness have become an open wound, one that won't heal and is easily aggravated, affecting every facet of my existence. "Lost" has become the refrain of my life since my husband's passing. I lost 20 pounds. I lost a sense of who I am. I lost my purpose. I lost my ability to focus, and even memories and the ability to complete simple tasks. They say grief can lead to "brain fog," and I find solace in knowing I am not losing my mind. It is especially excruciating to dispense copies of the death certificate, when tackling the paperwork that follows his passing. Sometimes as I stare at my own hands holding those documents, it feels like I am living in someone else's body or navigating a surreal world apart from my own.

ESTABLISHING CLEAR INTENTIONS FOR THE FUTURE: ALIGNING EXPECTATIONS BETWEEN BENEFACTORS AND RECIPIENTS

A well-structured legacy plan can serve as a reassuring and comforting road map during what can be a traumatic and confusing time for your loved ones. By providing clear guidelines, your plan empowers them to navigate difficult decisions with clarity and purpose. It alleviates the burden of uncertainty when they are already grappling with emotional challenges, ensuring they can focus on cherishing memories rather than being overwhelmed by complex logistical concerns. Additionally, a proactive legacy plan allows your family members to honor your wishes accurately, creating a lasting tribute to your values and beliefs. This thoughtful approach ensures that your principles are upheld in every decision they make, thereby reinforcing the family bonds you hold dear. Incorporating elements such as instructions for managing family heirlooms or charitable contributions can further personalize this legacy and enrich the connection between your family members and your memory.

It is important to remember that a legacy plan prepares for your own future as well as the well-being of your loved ones. As we stated at the beginning of this chapter, the plan is a gift and serves as an act of love, not only safeguarding your interests but also looking out for the emotional and financial stability of those you left behind. Therefore, promoting a sense of shared purpose from the beginning is a must. You can involve other family members in the discussion of your legacy plan to foster open communication, which can strengthen familial ties. Below are strategies to keep in mind.

1. Schedule purposive family meetings/gatherings. If needed, leverage technology to conduct virtual meetings.
2. Create a safe and open environment where young and old can share stories and express thoughts, concerns, and aspirations.
3. Encourage each family member to create a personal legacy statement that outlines their values, goals, and what they wish to pass on. This can create opportunities for honest discussions and help you to consider your family members' own ambitions for the future.
4. Share your reasoning behind your legacy decisions openly to build trust and understanding among family members, but remain flexible and willing to revisit and/or revise the plans to reflect current realities and expectations.
5. Create shared experiences. You might want to start including younger generations in some of your hobbies or projects that reflect the family values and legacy. Form a mentorship if necessary. This shared experience might later lead to acceptance and understanding of your legacy plan.

By involving other family members in the discussion of your legacy plan, you create a sense of shared purpose from the beginning and strengthen familial ties. Before we start to discuss the steps for creating purposes and goal setting in legacy planning, it is important for you to consider what you value in leaving a legacy and how you choose to communicate during legacy planning.

WHAT ARE YOUR LEGACY VALUES?

Harris (2023) wrote, "Values are like a compass, providing direction, while goals are the destinations we aim for." In the first section of the chapter, we provided you with tools to use for identifying your legacy's next stop on its journey. Now your values will, in large part, help you ensure that your legacy arrives at its destination! In this section, we briefly discuss some of the many meanings for *legacy* and value systems and provide activities to assist you with identifying your legacy values and how you may communicate these values, which will help you articulate where you want your legacy to go and how you want to get it there.

The word "legacy" is likely to evoke many different meanings for people. For example, for sports fans, legacy might characterize a sports team's impact, such as national titles or records of athletic achievement, and the positive effects of those achievements on its city's reputation and fans' experiences (Wu et al. 2021). To others, *legacy* may refer to one's effects on culture over time, such as an author's, musician's, artist's, or even a social influencer's body of work. And then, of course, there is one's personal legacy—the lasting impact an individual has on others through their actions, values, and contributions (González Rueda 2017). Lastly, and the focus of this chapter, others define legacy as a gift of one's property, such as an estate or other asset.

Sometimes, benefactors and beneficiaries such as institutions and family members agree as to the quality, meaning, and impact of a legacy and at other times they may disagree, further complicating the meanings and ultimately the "survival" of the legacy. To benefactors, a legacy might symbolize tradition, wealth, success, struggle, hard work, or sacrifice. Dissimilarly, legacy might symbolize conflict, burden, or obligation to the beneficiaries. A person's values will likely influence how one sees the act of giving a legacy.

One of the purposes of *Families in Ranching Succession Toolkit (FIRST)* in general and this chapter in particular is to provide knowledge and skills to the reader for the effective communication of goals during the many phases of legacy planning. The communication processes involved when gifting an asset can follow both formal channels (such as meetings with financial representatives, government agencies, and attorneys) and informal channels (such as discussions during family dinners, holidays, and other social events). Sometimes, unfortunately, there might be little to no communication between the asset owner(s) and the prospective recipients until it is too late to plan.

The process will require you to make many decisions that include but are not limited to legal considerations, familial considerations, and financial considerations. All considerations share at least one commonality: the need for effective communication for successful outcomes. What and how you communicate about your intentions—goals—to leave an asset as a legacy will be determined by what you value. So, what do you value about leaving a legacy? Knowing the answer to this question is a key first step in determining how you choose to communicate during legacy planning.

In general, values are standards or principles we use to help determine what is important to us. For many people, values may be both fixed and flexible—somewhat stable throughout our lives as well as likely to change over time due to personal circumstances and other factors internal and external to the individual. Also, some values may be both explicit or obvious and implicit or hidden to us and others. It is this dynamic of our values systems that calls us to examine from time to time what is of greatest consequence, which may provide clarity of expectations and hopefully result in feelings of satisfaction and fulfillment for those involved, especially when making a major decision to give a legacy.

Questionnaires can help people identify and communicate their values. A type of questionnaire known generally as an "adjective checklist" (Gough and Heilbrun 1965; Rocklage and Fazio 2015) has helped people further identify and understand their value systems in particular contexts such as that of legacy planning. Below, Activity 1 is intended to help bring more clarity to one's values about giving a legacy as well as support decision-making and improve communication about legacy planning itself.

ACTIVITY 1: WHAT DO YOU VALUE ABOUT GIVING YOUR LEGACY?

Step One. Read through the list of words below. Then, circle the words that best reflect your values about leaving your legacy. This step may be completed by legacy recipients, too. There is no minimum or maximum number of words you may select. If there are words that describe your values about giving a legacy that are not included in the list below, add these words to your list.

Activity Box 1

Enduring	Passionate	Trust	Ancestral	Equity
Generational	Balanced	Memory	Monetary	Family
Visionary	Purposeful	Vision	Sovereign	Bonds
Lasting	Relational	Founding	Profitable	Tradition
Resilient	Established	Foundation	Thriving	Inheritance
Ethical	Nurturing	Growth	Durable	Property
Familial	Ethical	Reciprocity	Invested	Agriculture
Charitable	Legacy	Inclusiveness	Equitable	Livelihood
Integral	Influence	Ideals	Inherited	Homestead
Influential	Tradition	Organization	Substantial	Community
Inspirational	Charity	Bonds	Expansive	Finance
Foundational	Morality	Integrity	Legal	Generations
Benevolent	Innovation	Stability	Steward-	Stewardship
Cultural	Value	Fulfillment	ship-based	Stability
Prosperous	Resilience	Mentorship	Culturally rich	Partnership
Collaborative	Preservation	Creativity	Organized	Harvest
Historical	Unity	Honesty	Resolute	Organization
Thoughtful	Altruism	Achievement	Homesteading	Culture
Loyal	Strategy	Reform	Wealthy	Compassion
Principled	Transparency	Abundant	Reputable	Heirloom
Transparent	Leadership	Rooted	Established	Sustainability
Unified	Wealth	Agricultural	Generous	Memory
Financial	Reputation	Fertile	Familial	Collaboration
Sustainable	Sustainability	Luxurious	Flourished	Extended family
Trustworthy	Collaboration	Bountiful	Balanced	Roots
Altruistic	Responsibility	Secure	Inspirational	Leadership
Diversified	Business	Generational	Nourished	Conservation
Structured	Family	Prosperous	Cultivated	Trust
Dynamic	Justice	Patriarchal	Meaningful	Growth
Moral	Harmony	Matriarchal	Worthy	Prosperity
Enduring	Law	Extended	Land	Resource
Generous	Commitment	Harmonious	Estate	Value
Traditional	Culture	Enriched	Acreage	Development
Wholesome	Prosperity	Rural	Farmstead	Planning
Pioneering	Connection	Historic	Wealth	Accountability
Devoted	Accountability	Cooperative	Money	Achievement
Innovative	Generosity		Security	

Step Two. Identify your themes or categories from the words you selected and indicate these in the space below. Themes should have a minimum of two associated words each, with no maximum number. As an example, if you selected the words Generational, Family, Tradition, and Roots, you might label this theme or category *Heritage*. Use a separate sheet of paper if the number of categories exceeds the available space.

Activity Box 2

Theme/Category	Associated Words
Heritage	Generational, Family, Tradition, Roots

The content and quantity of the themes you have identified in Steps One and Two of Activity 1 can be related to communication strategies—how and what you communicate from time to time—during your legacy planning journey. Now that you have selected your words and organized these words into themes or categories, count the number of themes you have identified. You might have only a few themes, but each is comprised of several words (or just a handful). Conversely, you may have many themes comprised of several (or just a few) words per theme.

The groundbreaking and long-standing work of many communication experts such as Jesse Delia and Barbara O'Keefe since the 1980s through the 1990s (Delia et al. 1982) and the more recent helpful summaries and applications by Griffin, Ledbetter, and Sparks (2019), Miller (2015), Littlejohn (1996), and many others have demonstrated how the content and quantity of categories and words per category can impact communication.[1] For example,

1. The authors have intentionally simplified and adapted the language of this theoretical explanation of Constructivism and its significant contribution to the communications field to support accessibility, engagement, and readability in the context of the accompanying activity. In doing so, some conceptual nuances and technical terminology have been omitted or

having more categories and words per category suggests your value system requires a complex and detailed approach to communicating about legacy giving. On the other hand, if you have identified relatively few themes and few words per theme, this can suggest less variation in your values around legacy giving; thus, you might be more likely to engage in more direct, basic, straightforward, and clear-cut communication approaches. The more diversified your list is, the more likely you will be able to adapt your communication approaches to the various aspects of the legacy context. For instance, you will be able to communicate effectively with many people of different positions, roles, and family relationships as well as adjust your communication based on the preferences of those to whom you will hand down your legacy. Or, as the researchers put it, your communication about your legacy is liable to be more "sophisticated" and more likely to help you achieve your goals.

This approach suggests there are three general message types related to the quantity and content of your themes, which we will paraphrase here for the purposes of this chapter: Self-expressive, Rules-based, and Strategic.[2] Self-expressive messages tend to be direct and can be emotional expressions of your feelings and thoughts about legacy planning. Rules-based messages tend to follow appropriate norms for interacting with others. Strategic messages adapt to the situation, such as when accommodations or compromises must be made to achieve goals. Keeping with our example category of Heritage, see Table 2.1 for effective and ineffective examples of each message type.

---

paraphrased. This modified version is designed to facilitate comprehension and application for learners, especially those who may be encountering the theory for the first time. While this simplification serves pedagogical goals, it does not reflect the full complexity or scope of the original theory. Readers interested in a deeper or more rigorous understanding are strongly encouraged to consult the primary sources and recommended scholarly works.

2. The original design logics—Expressive, Conventional, and Rhetorical—were developed by B. J. O'Keefe (1988) to describe increasingly sophisticated ways individuals construct messages based on their understanding of communication. For the purposes of accessibility in this context, these terms have been renamed as Self-expressive, Rules-based, and Strategic, respectively. These new terms are intended to more intuitively convey the practical orientation of each logic for lay audiences: Self-expressive reflects unfiltered, emotion-driven expression; Rules captures the reliance on social norms and communicative conventions; and Strategic emphasizes strategic, goal-oriented communication aimed at shaping relationships or outcomes. This renaming follows a broader trend in applied and instructional communication settings, where scholars and practitioners often adapt theoretical terminology for clarity and engagement (e.g., Tracy, Jackson, and Marin 1994).

Table 2.1. Message Types: Heritage Values

Message	Description	Effective	Ineffective
Self-expressive	Emotional expression of thoughts and feelings	"I feel really proud of what Grandpa and Grandma built for our family—it makes me want to be sure their hard work and dedication endure."	"Don't you care about our family roots? Why don't you want to continue the tradition? It would break their hearts."
Rules-based	Follows expectations for norms and appropriate behaviors	"We should do what we can to ensure their legacy and protect our heritage. We owe it to those who came before us."	"We've always managed this the way they wanted, so stop trying to change things."
Strategic	Adaptive and relational; seeks mutual understanding and good outcomes for all involved	"Let's find a way to share our heritage in ways that work for everyone so that their hard work may be enjoyed by generations to come and make sense given today's challenges."	"We can't accommodate everyone's interests; if family really cared about this, they would do things the way they've always been done"

As you can see from the table, value categories or themes can influence the content and relationship aspects of communication about a specific goal. In this case, the goal is to leave a legacy from a value perspective, where the legacy represents your heritage. Consequently, your communication about the legacy will reflect communication content and styles consistent with that value position. Using this approach will help you improve your communication, transitioning from simple self-expression to a strategic, relational approach. For example, when discussing legacy as heritage, the message styles range from the emotional to the strategic. In this case, given the legacy value of heritage, the effective strategic approach will result in others' feeling invited into the planning process. Growing your awareness of these values-related message styles will promote greater empathy and intentional communication throughout the planning process.

WHEN LEGACY PLANNING GOES WELL: THE BRIGHT SIDE OF LEGACY PLANNING

When legacy planning goes well, future generations that own or operate the land are equipped for success. When the estate plan has been thoughtfully crafted with the heirs' abilities, goals, and resources in mind, success is much more likely. Generally speaking, farmers and ranchers want to leave their business to their children in anticipation of the children's having much greater success than their predecessors. However, a plan must be in place and communicated in order for that to happen. See the example story about the family below and the success the heirs enjoyed because of a well-thought-out and well-communicated plan.

> *Jack was the child of sharecroppers. After watching his parents struggle financially and physically to make a meager living, he dreamed of having a farm of his own. After Jack returned home from Europe following World War II, he married Maxine. With a little extra money in his pocket that he had saved during the war, he and Maxine made a down payment on a 160-acre farm in 1952. They farmed a variety of grain crops, had a few cattle, and made a good living. They left a strong agribusiness to their two boys, Jerry and Russell. Jerry and Russell each got married, and each had three children. They continued to farm together, purchased more land, and diversified their operation until their children were adults.*
>
> *In the beginning, Jerry and Russell hoped their children would all farm together. Then, two things happened. First, a nearby urban area began to encroach upon their farm. Second, with six different heirs, each with different levels of interest in the farm, Jerry and Russell realized the farm would not support six more families. They knew farming traditional grain crops would become increasingly difficult as a major highway was being built adjacent to their land, a neighborhood was being developed nearby, and their local grain elevators had closed.*
>
> *The brothers assembled their adult children and discussed a nontraditional way to keep the farm in production: agritourism. The family decided to plant a corn maze, pumpkin patch, apple orchard, and farm store. They agreed to give it two years and reevaluate the situation at the end of that time.*
>
> *The plan worked incredibly well. The adult children offered ideas to expand the operation and appeal to more consumers. Each family*

member's unique talents and skills contributed to the farm, and each wanted to participate. Jerry and Russell saw the potential, so they set up a corporation so they could run the farm like a business. Each child was a shareholder and employee of the business.

When Jerry's and Russell's grandchildren grew into adulthood, another generation of the family wanted to be involved in the farm. During a family meeting, Jerry, Russell, and their adult children created a new set of goals: Keep the farm in production, delight and educate the public with a quality agritourism experience, and provide a full- or part-time job for any family member who wanted to remain on the farm. They also agreed that after ten years of working for the farm, the grandchildren could become shareholders in the business.

One of the grandchildren has a talent for marketing and oversees all advertising and media relations. Another has a head for business and manages all accounts and finances. Yet another grandkid is the plant person, and she handles all planting, maintenance, and harvesting. The other grandkids contribute in different, valuable ways. Decisions about the farm are made collectively at regular business meetings.

Because Jerry and Russell communicated with their kids, then incorporated the farm, they were able to pass the operation on to the next generation and the one after that. Their adult kids saw a good example of how a farm could remain intact and stay in the family.

The key reason for the success in Jerry's and Russell's story was that they communicated with their potential heirs. They learned what the adult children wanted to do and how to contribute, then the older generation created resources so the younger family members could be successful and stay on the farm. Each person's skills were taken into account so they could contribute to the business, and if a grandchild chose to find employment elsewhere, that was okay.

Owning and running a family agribusiness is not easy. There will be disagreements, differences of opinions, hurt feelings, and squabbles from time to time. However, when the goal is clearly stated to all involved—with input from everyone—it is so much easier for everyone to know what they are striving toward. Working through problems can help strengthen relationships within the family and the business.

If Jerry and Russell had not stated their goal—to see their children grow up and farm together—their children would not have put so much effort into changing the business model. The children realized that a clear goal made decision-making much easier, so that when the time came, they created a new set of goals to move the family and business forward for the next generation of owners.

Research shows that only 12 percent of small businesses survive into the third generation. However, with clear, realistic goals, passing a farm or ranch on to the next generation is much more attainable. When the goals have been met, new goals can be written for the next generation to work toward.

WHEN LEGACY PLANNING DOES NOT GO WELL: THE DARK SIDE OF LEGACY PLANNING

> *Tim and Janet have three children and 2,000 acres of farmland. Knowing they wanted to leave each child a farm, they utilized the services of an estate planning attorney. They were instructed that the "share and share alike" strategy rarely works, so they thought it best to leave specific plots of land to each child. Also, they stipulated in their will that their land was to remain in production agriculture. Years before, Tim and Janet saw the neighbors' children squabbling over their inheritance, so they worded their will to revoke a child's inheritance if any of the heirs negotiated a different piece of land, bickered, or fought about the land they received.*
>
> *Tim and Janet had good intentions: They wanted their children to appreciate the land they were gifted, and they wanted their kids to act civilly to one another. However, this ironclad will created problems. A wind energy company approached the heirs—their land is in the perfect spot for a wind farm. The money offered is significant; they would be foolish to refuse. In addition, the heirs would like to swap some of their land around to make a more even distribution of the windmills, and thus the profits. The adult children are agreeable with one another and want to honor their parents' wishes, but the will is so inflexible they are not able to use the land in the way they see fit.*
>
> *The heirs do not want to disrespect their parents; however, it would take fifteen years of great agricultural production to yield the amount of money offered by the wind company.*

The story of Tim and Janet is what the authors of this chapter call "ruling from the grave." It does not account for the heirs' wants, needs, or goals.

Overly restrictive inheritance conditions can lead to resentment toward the predecessors, spark conflicts, and possibly create greater financial constraints for the heirs.

When crafting a succession plan, talk to your heirs to learn about *their* goals and current situation. Although leaving land is a wonderful gift, some people do not have the capability to manage a ranch. Consider where the heirs live, their knowledge level of production agriculture, their current financial situation and family structure. Some people have no interest and/or knowledge in owning land, and a ranch, especially with stringent restrictions, could be burdensome. In a strict will like that mentioned in the story above, the parents, even though acting with the best intentions, do not allow their children the flexibility to maximize the inheritance potential.

It is perfectly fine for a parent or grandparent to communicate their wishes for their estate to their children and grandchildren. However, it's not okay to saddle them with instructions that will fill them with guilt if they are unable to carry them out, as is the case in the story below.

> *Irene had five children and nine grandchildren. In her 91-year lifetime, she was able to amass an estate that included 10,000 acres of ranch land.*
>
> *When Irene died, her land and money went into a trust, and her children hired a ranch manager. The arrangement worked fine, until two of Irene's children died and the grandchildren inherited a percentage of the trust. Several grandchildren wanted to dissolve the trust, sell the land, and divide the assets. One vocal granddaughter was adamantly opposed to this idea, telling her siblings and cousins, "We cannot sell this land. Grandma didn't want that. She wanted us all to farm it together." Several grandchildren had never been told their grandmother's wishes; others knew their grandmother wanted them to keep the land but were aware of the impracticality of her wishes.*
>
> *In the end, the heirs dissolved the estate, sold the land, and split the money. The granddaughter who was especially vocal is somewhat estranged from the family and rarely speaks to her siblings or cousins anymore.*

We can't know the conversations Irene had with her granddaughter, but the latter was convinced that her grandmother wanted all the land kept together. However, although Irene worked very hard to build her estate, retaining the land in its entirety might not be practical, especially when multiple generations of heirs are owners. In the end, the lack of communication created hard feelings and fractured the family, which no parent wants. Ideally, Irene should have made sure to discuss her wishes in person with *all* the heirs to determine if keeping a ranch that size in one piece would be practical for her heirs, then communicate her final decision to the whole group to minimize misunderstandings. Instead, her ranch was sold, and some members of her family won't speak to one another.

The story below is similar in that an heir felt so compelled to honor his predecessors' legacy.

> *Will was very close with his grandfather, Henry. Will was the only child and grandchild, so he was raised to love the land and farm it so that he could leave it to his children or grandchildren.*
>
> *Will worked very hard all his life to keep and grow the farm that was gifted to him; however, after five consecutive years of low rainfall, blistering heat, and low commodity prices, he is looking at a bleak financial future. Will's children have encouraged him to sell the farm, but he refuses. "My granddad gave me that land, and I will die trying to keep it," he has said on more than one occasion. But that's the problem. His children fear he may damage his health or perhaps even die from trying to keep the farm operational to honor a decades-long emotional attachment.*

Legacy attachments to ranches are very real. Many farm and ranch kids were raised to love the land, appreciate its value, and keep land for the next generation. However, sometimes it simply isn't realistic to do so. As illustrated in the story above, the struggle to keep the land in the family was damaging Will's physical and mental health. A conversation between Henry and his grandson about the gift of land and how to use it as he needed would have lightened the emotional load that Will was carrying for himself and his family

GOAL SETTING IN LEGACY PLANNING

To get an overall picture of the preferences of all involved in a potential legacy gift, use each member's statements, the value exercises mentioned above, and/or stories to identify their major goals. Then, categorize those goals into distinct categories. Finally, use a Venn diagram with circles representing each category of goals. The Venn diagram will show overlaps and differences.

Through the exercises and discussions offered above, common goals will emerge and be identified. After a shared vision has been determined, you can delineate achievable outcomes. Then, set specific, measurable, achievable, relevant, time-bound (SMART) goals to track progress and expectations.

CONCLUSION

There are lots of ways to learn more about crafting an estate plan. First, look for workshops or seminars in your area. The Cooperative Extension Service, agricultural organizations, government agencies, and even some houses of worship offer short seminars, webinars, and workshops to help you create your estate plan; many are very reasonably priced or even free. Attorneys, tax professionals, banks, and insurance agents can be a great source of information, and again, Cooperative Extension in some states will have an expert who specializes in estate planning. Finally, those family meetings are hugely important to learn more about your heirs' wishes and to communicate your goals with them.

CHAPTER 3

Budgeting, Financial Planning, Estate Planning, and Taxation

Kelly Lange and Mitzi Lauderdale

SUCCESSION IN RANCHING CONTEXT

Financial and legal planning are essential to aiding a smooth transition and ensuring that assets are passed on as intended, without disruption. Budgeting, financial planning, estate planning, taxation, and an understanding of intestate succession laws are key to effective transfer planning.

Succession planning and estate planning are not necessarily the same thing, although there is considerable overlap in the processes. In general, succession planning is often used to describe the transfer of management responsibilities from one generation to the next. These responsibilities include day-to-day and long-term decision-making, personnel management decisions, and financial management decisions, including financing arrangements and bill-paying responsibilities (Gasson and Errington 1993). The pace at which management responsibilities are transferred from one generation to the next varies considerably, and it is not uncommon for the complete transfer process to take many years.

The estate planning and inheritance process generally focuses on the transfer of business ownership, which may or may not be closely integrated with the transfer of management responsibilities (Gasson and Errington 1993). Similar to the transfer of management, the ownership transfer process will vary by family, and the transfer may happen all at once or bit by bit over time.

BUDGETING FOR SUCCESSION: UNDERSTANDING THE NUMBERS

Several steps must be taken to determine whether the business can financially support multiple generations. Budgeting for operations, investment, and inheritance- and estate-related expenses must be considered. Cash flows must be budgeted and managed to accommodate long-term transfer goals, particularly during the short-term transition. Additionally, valuation records for large balance sheet assets, such as land, equipment, and livestock, must be kept up to date and maintained accurately (Barry and Ellinger 2010; Kay, Edwards, and Duffy 2024).

To determine if the business is large enough to realistically support multiple generations, a thorough assessment of the business legal and financial status must be conducted. Begin with basic information relative to the business.

Table 3.1. Basic Information Relative to the Business

Size of operation (in acres)	
Number of acres owned	
Number of acres leased	
Location of operation (list all locations)	
Legal structure of business (e.g., proprietorship, legal partnership, LLC, corporation, etc.)	
Principal operator(s)	
Potential successors/heirs	
Primary production / commodities produced	
Financial basis of owned acreage	
Current estimated fair market value of owned acreage	

Next, basic financial statements need to be developed to assess the business financial position. Begin by evaluating assets, liabilities, and owner equity (Barry and Ellinger 2010).

Table 3.2. Statement of Financial Position

Name of business: **As of (date):**	
ASSETS	
Current Assets	
Cash / savings	
Accounts receivable	
Inventories (market livestock, crops, supplies, etc.)	
Prepaid expenses	
Other current farm assets	
Other current personal assets	
Total Current Assets	
Noncurrent Assets	
Breeding livestock (market value)	
Machinery / equipment	
Capital leases	
Investment in other assets	
Retirement accounts	
Life insurance (cash value)	
Business real estate	
Land	

Buildings / structures / facilities / improvements	
Other noncurrent farm assets	
Other noncurrent personal assets	
Total Noncurrent Assets	
Total Assets	
LIABILITIES	
Current Liabilities	
Accounts payable	
Current notes payable	
Current portion of noncurrent debt	
Accrued interest	
Accrued expenses (taxes, rents, etc.)	
Current deferred taxes	
Other current business liabilities	
Other current personal liabilities	
Total Current Liabilities	
Noncurrent Liabilities	
Noncurrent debt	
Real estate debt	
Capital leases	

Noncurrent deferred taxes	
Other noncurrent business liabilities	
Other noncurrent personal liabilities	
Total Noncurrent Liabilities	
OWNER EQUITY	
Contributed capital	
Retained earnings	
Market gain	
Equity in personal assets	
Total Owner Equity	
Total Liabilities and Owner Equity	

When assessing the business financial position, asset valuation relative to cost and depreciation should also be considered (Barry and Ellinger 2010).

Table 3.3. Asset Valuation Relative to Cost and Depreciation

Asset	**Original Cost**	**Accumulated Depreciation**	**Book Value (cost less depreciation)**
Breeding livestock			
Machinery / equipment			
Land			
Building / facilities / improvement			

After accounting for all business assets, liabilities, and owner equity, construct an income statement to evaluate business revenues and expenses (Barry and Ellinger 2010).

Table 3.4. Income Statement

Name of business: **As of (date):**	
REVENUE	
Livestock receipts	
Crop receipts	
Inventory adjustments	
Cash sales – other	
Change in value of raised breeding livestock	
Gain / loss culled livestock	
Changes in accounts receivable	
Government programs / insurance payments	
Other income	
Gross Revenue	
EXPENSES	
Livestock expenses	
Crop expenses	
Inventory adjustments	
Cash expenses – other	
Repairs / maintenance	
Insurance	
Land rent	
Utilities	
Depreciation expense	
Adjustments to expenses	
Taxes	
Total Operating Expenses	
Income from Operations	

Interest expense	
Gain / loss on sales of capital assets	
Net Farm Income	

A statement of cash flows is a financial statement that documents all cash transactions, including operating, financing, and investment activities, for the business. The cash flow statement is similar to an income statement in some regards, but there are distinct differences, as the statement of cash flows documents all cash inflows and outflows, including nonbusiness-related cash flows, that impact the cash position of the business (Barry and Ellinger 2010).

Table 3.5. Statement of Cash Flows

Name of business: **For the 12-month time period (dates):**	
Cash Flows from Operating Activities	
Inflows	
Cash from production	
Cash from government payments / insurance	
Cash from nonbusiness income	
Total Cash Inflows from Operating Activities	
Outflows	
Cash paid for operating expenses	
Cash paid for interest expenses	
Cash paid for market livestock	
Cash paid for taxes	
Cash paid for family living expenses	
Total Cash Outflows from Operating Activities	
Net Cash Flows from Operating Activities	
Cash Flows from Investing Activities	
Inflows	
Cash received from sales of machinery / equipment	
Cash received from sales of real estate	

Cash received from sales of breeding livestock	
Cash received from savings, personal assets, etc.	
Total Cash Inflows from Investing Activities	
Outflows	
Cash paid to purchase machinery / equipment	
Cash paid to purchase real estate	
Cash paid to purchase breeding livestock	
Cash paid / deposited for savings, personal assets, etc.	
Total Cash Outflows from Investing Activities	
Net Cash Flows from Investing Activities	
Cash Flows from Financing Activities	
Inflows	
Cash from operating and short-term loans	
Cash from long-term loans	
Cash from capital contributions	
Total Cash Inflows from Financing Activities	
Outflows	
Cash paid for operating and short-term loans	
Cash paid for long term loans (scheduled)	
Cash paid for long term loan (unscheduled)	
Cash paid for capital leases	
Cash paid for capital contributions	
Total Cash Outflows from Financing Activities	
Net Increase (Decrease) in Cash Flows	
Beginning-of-Year Cash Balance	
End-of-Year Cash Balance	

FINANCIAL PLANNING FOR THE FUTURE: FOSTERING LONG-TERM STABILITY

Financial strategies should be aligned with the long-term financial vision for the business. Planning for risk and uncertainty is vital for the protection of family assets. Additionally, use of financial leverage should be evaluated and debt management strategies considered during intergenerational transfers. This often includes consideration of financing strategies to retain ownership and avoid the need for sudden asset sales to generate cash (Kay, Edwards, and Duffy 2024).

LONG-TERM FINANCIAL VISION

Consider what you would like your business to look like in the future. A goal for many owners is to facilitate strong financial performance and make good investment decisions. Additionally, for many business owners, an increase in equity and a decreased reliance on debt financing are desired goals. Write down some of your future long-term financial visions and goals for your business.

Table 3.6. Long-Term Financial Visions and Goals for Your Business

In five years, I would like my business to:
In ten years, I would like my business to:
In fifteen years, I would like my business to:
In twenty years, I would like my business to:

RISK MANAGEMENT

Risk is an inherent part of life. While it may not be possible to completely eliminate risk, there are many effective ways to mitigate risk. Agricultural production is subject to a variety of business risks that may impact the financial stability of the business. Financial risk must also be considered if the business utilizes loaned funds. It is important to keep in mind that as the level of debt financing increases, the impact of these business risks will be amplified.

For example, a highly leveraged business with significant loans to repay will struggle more financially during a bad production year or market downturn compared to a less leveraged business (Kay, Edwards, and Duffy 2024).

The following table lists various business risks experienced by agricultural producers. Consider each category of risk and indicate whether your business is adequately mitigating these risks and which areas might need to be addressed.

Table 3.7. Business Risks Experienced by Agricultural Producers

Category of Risk	**Risk Related to:**	**Am I adequately addressing these risks?**	**What could I do to more effectively address these risks?**
Production, yield, and technological risks	Weather, disease, infestation, new technology adoption and obsolescence, etc.		
Market risks	Supply and demand for inputs and outputs		

Legal, human resources, and policy risks	Dangers associated with agricultural production, livestock, and machinery. Also includes human resources management issues, tax laws, government programs, etc.		
Personal risks	Potential for unexpected illness, injury, family disputes, succession and estate planning, personal financial issues, etc.		

Now consider how financial risk, which is related to borrowing and interest rates, might impact or increase these business risks.

Table 3.8. Analysis of the Impact of Financial Risk

I would consider my current leverage position (amount of borrowed funds) to be: Low Medium High	List all notes payable by the business here. Be sure to include total amount of each loan, loan term to maturity, repayment schedule (monthly, quarterly, annual, etc.), payment amount, and interest rate.

I feel like I have a plan in place to ensure I pay off my debts on time, with minimal interest paid, and have a contingency plan in place in case of repayment difficulty: Yes No I'm not sure	What are some strategies I could consider to minimize my borrowing needs and/or make paying off my debt obligations easier?
How might my leverage position be impacting other categories of business risk? Do I need to address some of these issues?	

ESTATE PLANNING: LEGAL STRUCTURES AND WEALTH TRANSFER

The impact of business legal structure should not be overlooked during the business transition. Options for the transfer of business ownership must be considered and key estate planning documents, such as wills or trusts, should be developed (Dalton and Langdon 2022). These types of documents will outline the legal considerations for assets to help aid a smooth estate transfer.

Impact of Legal Structures on Business Transition

An overview of common ways to transfer business ownership and questions to help evaluate your best path.

The way your business is legally structured has significant implications for how ownership is transferred, how the business continues after your exit, and how assets are treated for tax and estate purposes. Some structures offer flexibility and protection, while others require more planning to avoid disruption during a transition.

Some common business structures include the following, each with unique implications for ownership transfer, tax treatment, and succession. In community property states like Texas, it is especially important to consider how marital property laws may affect control, valuation, and transferability—particularly when interests were acquired during marriage.

Table 3.9. Common Business Structures and Key Considerations

Business Structure	Benefits and Challenges	Community Property Considerations
Sole Proprietorship	Benefits: Simple to operate, minimal cost Challenges: No legal distinction between owner and business; dissolves upon owner's death unless planned for	Business may be considered community property; spouse may have rights even if not involved in business
General Partnership	Benefits: Shared management and profits Challenges: Requires formal agreement to transfer ownership; risk of dissolution	Spousal interest may exist in partner's share; formal consent or agreement may be needed
Limited Liability Company	Benefits: Flexible structure; liability protection Challenges: Transfer may require member approval	Member interests acquired during marriage may be community property; transfer or sale can be complex
S Corporation (S-Corp)	Benefits: Liability protection; avoids double taxation Challenges: Limited shareholders; risk of disqualification on improper transfer	Spouse may count as a shareholder; agreement needed to manage or transfer ownership
C Corporation (C-Corp)	Benefits: Easier share transfer; suitable for complex structures Challenges: Double taxation; more regulatory formalities	Shares acquired during marriage are typically community property; must account for in succession

Family Limited Partnership	Benefits: Enables family ownership transfer; control remains with senior generation Challenges: Requires legal setup and IRS compliance	Partner interests may be partially owned by spouses; can affect valuation, control, and tax outcomes

Reflection Questions. Before moving forward, take a moment to evaluate your current business framework and how it aligns with your long-term transition goals.

1. What is the current legal structure of your business (or a business you advise)?
2. How might that structure impact the ease or complexity of transferring ownership?
3. Are there existing legal agreements or documents that clarify ownership rights and responsibilities?
4. Would a change in structure—such as forming an LLC or FLP—better support your goals for control, tax planning, or intergenerational transfer?
5. If your business is family-owned, would a Family Limited Partnership offer advantages worth exploring with legal and tax advisors? See the tax section for more information.

Note: Revisiting legal structure is not just a legal or tax exercise—it's a strategic decision that shapes long-term continuity, family dynamics, and legacy.

Ownership Transfer Methods

What to know and consider before choosing how to hand over your business.

When you're ready to transition your business, you'll need to decide how ownership will be transferred. Each method has financial, tax, and control implications. Before meeting with an attorney or advisor, take time to reflect on which options might fit your goals.

Table 3.10. Business Ownership Transfer Methods

Transfer Option	Description
Gifting	Transfer ownership to family over time. Often used to reduce estate size and allow gradual succession.
Sale to Family	Ownership is sold to a family member, allowing you to receive payment. Can be structured with flexibility.
Third-Party Sale	Sell the business outright to an external buyer. Common when no internal successor is identified.
ESOP (Employee Ownership)	Ownership is transferred to employees via a trust. Helps retain culture and reward staff loyalty.
Buy-Sell Agreement	A legal agreement (often among partners or family) outlining how and when ownership changes hands.

Key Estate Planning Documents

Key documents that support business succession and identify what to consider before meeting with your attorney.

A solid estate plan ensures your business and other assets transfer according to your wishes—and with minimal confusion or delay. Certain documents are especially important for business owners preparing for succession.

Table 3.11. Essential Estate Planning Documents for Business Owners

Document	Description
Will	Outlines who inherits your assets—including your business—after death.
Living Trust	Enables assets to transfer outside of probate; can also provide management continuity if incapacitated.
Durable Power of Attorney	Authorizes someone to manage financial and legal matters if you are unable to do so.
Buy-Sell Agreement	Specifies what happens to business ownership if you or a partner retire, become disabled, or die.
Business Succession Plan	A road map identifying who will take over the business, how the transfer will occur, and when.

Legal Considerations for Assets

A guide to evaluating how your business and personal assets are titled, valued, and prepared for transfer.

Smooth transitions require more than just a succession plan—they also depend on how business and personal assets are structured. If assets are not clearly titled, valued, or assigned, they can become roadblocks during estate settlement or business transfer. Legal clarity helps ensure your intentions are honored without delay, dispute, or tax complications.

Table 3.12. Asset Structuring for Seamless Business and Estate Transitions

Key Consideration	Description
Asset Ownership	Who legally owns each asset? Is it in your name, the business's name, or held in trust? Ownership impacts how (and whether) it can transfer.
Business Valuation	Without a current valuation, you may misjudge your business's worth, leading to tax issues or false assumptions during succession planning.
Beneficiary Designations	Some assets (e.g., retirement plans, insurance) bypass your will and transfer directly to listed beneficiaries.
Ownership Disputes or Complexities	Lack of documentation or conflicting claims among heirs, family, or partners can delay or prevent a successful transition.

How an asset is titled—or legally owned—can significantly impact how it is transferred at death, whether it passes through probate, and who ultimately receives it. In community property states like Texas, titling becomes even more important, as marital property laws may affect business and personal asset distribution. Table 3.13 outlines common titling options, who holds ownership, and what each means for succession and estate planning.

Table 3.13. Asset Titling and Succession

Title Type	Who Owns It	Key Features	Succession Impact
Sole Ownership	One person	Owner has full control; asset passes via will or probate	Included in estate; must be transferred through will or trust
Joint Tenancy with Right of Survivorship	Two or more individuals	Equal ownership; automatic transfer to survivor(s) upon death	Avoids probate; not controlled by will
Tenancy in Common	Two or more individuals	Each owner can have unequal shares; interest passes via will or estate	Must go through probate for deceased's share

Community Property (in states like TX)	Married spouses	Equal ownership of assets acquired during marriage (unless otherwise agreed)	Each spouse may direct their half via will; subject to probate
Community Property with Right of Survivorship (CPROS)	Married spouses (in states like TX)	Combines equal ownership with automatic survivorship; avoids probate	Surviving spouse automatically inherits entire asset; avoids probate
Tenancy by the Entirety (not in TX)	Married spouses	Similar to joint tenancy but with stronger creditor protections (in some states)	Survivor receives full ownership; bypasses probate
Revocable Living Trust	Trust (controlled by grantor)	Assets owned by trust; avoids probate; can provide management if incapacitated	Transfers under trust terms; avoids probate
Entity Ownership (e.g., LLC, Corporation)	Business entity	Legal entity owns the asset; governed by operating or shareholder agreements	Transfer guided by entity documents; may avoid probate and allow continuity

Structuring Estates for Smooth Transfer

How to prepare your estate so business and personal assets transfer clearly, efficiently, and in line with your wishes.

Estate planning is not just about personal assets. For business owners, it's also about ensuring that ownership, control, and value of the business pass smoothly to the next generation—or to your chosen successor. Without proper planning, even a successful business can become legally entangled in probate, tax liability, or family disputes. To avoid this predicament, you need a coordinated estate plan that clearly defines:

- What you own (and how it's titled)
- Who receives it
- How it will transfer
- When it will transfer (during your lifetime or after death).

Table 3.14. Core Estate Planning Tools

Tool	Purpose
Will	Specifies how your personal and business assets are distributed after death; names guardians and executors.
Revocable Living Trust	Avoids probate and allows faster private asset transfer; you retain control during your lifetime.
Durable Power of Attorney	Appoints someone to manage financial and legal matters if you become incapacitated.
Advance Directive	Outlines your preferences for medical care if you are unable to communicate them.
Buy-Sell Agreement	Defines ownership transfer terms if an owner dies, retires, or becomes disabled—typically used among partners or family.
Beneficiary Designations	Assigns asset recipients for accounts like insurance or retirement plans; overrides directions in your will or trust.

ACTIVITY: REVIEW YOUR ESTATE PLANNING READINESS

Answer these prompts to identify strengths and gaps in your estate planning:

1. Which of the following documents have you already created?
    a. Will
    b. Revocable Living Trust
    c. Durable Power of Attorney
    d. Advance Directive
    e. Buy-Sell Agreement
    f. None or unsure

2. When were these last reviewed or updated?
    a. Within the last three years
    b. More than three years ago
    c. Not sure

3. Do you know how your business ownership is titled and what happens to it upon your death?
    a. Yes, it's clearly outlined
    b. Partially—I need to review
    c. No

4. Have you discussed your estate plans with key people (e.g., family, successor, partner)?
    a. Yes, fully
    b. Somewhat
    c. No

Table 3.15. Common Pitfalls to Avoid

Common Pitfall	Explanation
Assuming Your Will Covers Everything	Some assets, such as retirement accounts and life insurance, transfer based on beneficiary designations—not your will.
Leaving Plans Vague	Lack of clarity in your estate or succession plan can lead to confusion, conflict, or legal disputes.
Failing to Coordinate Business and Personal Planning	Your business may be your largest asset and must be intentionally included in your overall estate strategy.
Delaying Communication	Key individuals (e.g., family, partners) need to understand your plans to ensure a smooth and supported transition.

FINAL SUMMARY WORKSHEET: PREPARING FOR YOUR ATTORNEY

Use the outline below to summarize what you've learned and identify key talking points for a professional estate planning conversation.

1. My preferred successor(s):
    a.
    b.

2. Business structure and ownership status:
    a.

3. Estate documents I currently have:
    a.
    b
    c.
    d.

4. Top concerns I want to address in planning:
    a.
    b.

5. Questions I want to ask my attorney:
    a
    b.

MULTIGENERATIONAL MANAGEMENT: ALLOCATING MANAGEMENT RESPONSIBILITY

Management allocation among multiple generations should be determined early in the transfer process. If decision-making will be shared between generations, it is important to denote who will be responsible for which types of decisions, particularly personnel management decisions and financial decisions. The family must be honest with each other and determine if management decisions can realistically be shared without conflict. Begin by identifying all individuals, whether family members or external stakeholders, who will be assuming a role within the business.

Table 3.16. Multigenerational Management Allocation and Decision-Making

Family Member and/or Stakeholder	Role in the Business

Once these individuals have been determined, consider the multigenerational managerial decision-making process. Will decision-making be shared among individuals? If so, who will be responsible for which decisions? What is the intended process for transferring management responsibilities from one generation to the next? Be frank about the sharing of management responsibilities between generations. Do you *realistically* believe these responsibilities can be shared without undue family conflict? If not, then alternative strategies should be considered, such as transferring management responsibilities concurrently with ownership of the business upon the death of the older generation.

Table 3.17. Structuring Management Responsibilities Across Generations

Type of Management Decision	Who is currently responsible for these decisions?	Who will be responsible for these decisions once business succession is complete?	Do you anticipate these responsibilities being shared for a period of time?	If decision-making might be shared between generations, what do you envision this looking like?
Day-to-day decisions				
Long-term (five- to ten-year+) decisions				
Personnel management decisions				
Borrowing and lending decisions				
Oversight of the business finances				

Miscellaneous decisions				

NAVIGATING TAXATION IN SUCCESSION: MINIMIZING THE TAX BURDEN

It is important to understand the tax implications of the succession process. These implications can be complex, particularly as regards gift, estate, inheritance, asset division, sales, and capital gains issues (Bost et al. 2016). Similarly, methods of mitigating taxes should be considered.

Succession planning isn't just about who takes over your business—it's also about how much of your business's value is retained for your family, heirs, or new owners. Taxes can take a significant bite out of your estate or the business's value if not carefully planned for.

Effective succession plans include strategies to minimize these burdens and structure transfers efficiently. The method you choose for ownership transfer—gifting, selling, or inheritance—has major tax implications. Table 3.18 provides a quick overview of common types of taxes related to succession.

Table 3.18. Tax Considerations for Business Transfer

Tax Type	What It Covers
Gift Tax	Applied when you transfer ownership without receiving full value in return. Annual exclusions apply.
Estate Tax	Imposed on the estate's value at death, above federal or state exemption thresholds.
Inheritance Tax	Paid by heirs in some states based on what they inherit. (Note: Texas currently does not have this tax.)
Capital Gains Tax	Applies to appreciated assets when they are sold. May be triggered if the business is sold during life.
Income Tax	May apply to structured payouts (e.g., installment sales or retirement plans).
Generation-Skipping Transfer Tax (GST Tax)	A federal tax applied to transfers made to beneficiaries who are two or more generations below the donor (e.g., grandchildren), above exemption limits.

Each transfer method comes with its own set of tax consequences. Use Table 3.19 to compare how different options might impact your succession goals.

Table 3.19. Tax Implications for Business Transfer Options

Transfer Method	Potential Tax Impact
Gifting	Can reduce estate size but may trigger gift tax above IRS exclusion limits.
Sale to Family	Provides income but may trigger capital gains and must meet fair market value rules.
Third-Party Sale	Likely triggers capital gains tax; may also include depreciation recapture.
ESOP	Offers favorable tax treatment for owners and employee trusts when structured correctly.
Inheritance	Often receives a "step-up in basis," reducing capital gains if sold shortly after inheritance.

The strategies listed in Table 3.20 are commonly used to reduce estate size, defer taxes, or structure business transfers more efficiently. These can be discussed with a tax advisor or estate planning attorney.

Table 3.20. Strategies to Mitigate Business Succession Taxes

Strategy	How It Helps
Annual Gift Exclusion	Allows tax-free gifts up to a set limit per person, per year (e.g., $18,000 in 2024).
Grantor Retained Annuity Trust (GRAT)	Transfers appreciation to heirs while retaining income for a period.
Family Limited Partnerships (FLPs)	Discounts value for tax purposes and retains control with senior family members.
Step-Up in Basis at Death	Heirs inherit assets at current market value, minimizing future capital gains.
Private Annuities	Transfers assets in exchange for unsecured lifetime payments; may avoid gift tax and reduce estate size.
Self-Canceling Installment Notes (SCINs)	Allows sale of an asset with installment payments that cancel at death, potentially excluding value from estate.

If supporting charitable causes is important to you, the giving strategies featured in Table 3.21 can also offer significant tax benefits. They allow for legacy planning while potentially reducing income, estate, and capital gains taxes.

Table 3.21. Charitable Giving Strategies for Tax Efficiency

Strategy	How It Helps
Charitable Gifts	Outright donations to qualified charities reduce income and estate taxes and support philanthropic goals.
Charitable Remainder Trusts (CRTs)	Transfer assets while retaining income; reduces estate and defers capital gains taxes.
Charitable Gift Annuities	Transfer assets to a charity in exchange for lifetime payments and receive a partial tax deduction.

The strategies outlined in this section represent commonly used tools for reducing the tax burden during succession. While these approaches offer valuable planning benefits for many business owners, more advanced strategies may be appropriate depending on the size, complexity, and nature of your estate.

Specialized planning options—such as Section 2032A Special Use Valuation for qualifying farms and ranches, generation-skipping trusts, and intentionally defective grantor trusts (IDGTs)—can offer additional tax advantages when properly structured. These techniques typically require the guidance of experienced estate and tax professionals.

REFLECTION QUESTIONS. ANSWER THESE QUESTIONS TO PREPARE FOR MEETING WITH A CPA OR ESTATE ATTORNEY.

1. What is the current market value of your business or key assets?
    a. Formal valuation completed
    b. Informal estimate only
    c. Not sure

2. Have you discussed potential capital gains or estate tax exposure with a professional?
    a. Yes
    b. No
    c. Planning to

3. Have you made annual gifts to reduce estate value?
    a. Yes
    b. No

4. Do you want to prioritize minimizing taxes, preserving control, or generating income from the transition?
5. Which tax strategies above are you most interested in exploring further?

As always, consult with your attorney, CPA, or estate planner to evaluate which combination of tools aligns best with your succession goals.

CHAPTER 4

Mediating Associated and Unseen Conflicts[1]

Jason Headrick and Gene Valentini

Ranching and farming require much time and hard work to yield a successful operation. Also necessary are the cooperation of the weather, the contributions of a variety of people, and many other variables. Because people are involved in the equation, conflict is bound to come up from time to time. Conflict is a normal function of relationships; how we address it affects the status of relationships and the operation itself. This chapter will provide a discussion on defining conflict, suggestions on how to appropriately manage conflict when it arises, an overview of mediation tactics, and examples for you to consider in your own work toward navigating conflict and succession planning. Individuals manage conflict differently: Some people pretend it is not happening; others participate but do not take it seriously; and still others thrive on it.

Regardless, we will face conflicts in our lives, and sometimes conflict will be associated with our ranches and farms.

CASE STUDY: THE NOT-SO-BRIGHT SIDE OF SOLAR PANELS

A family's plans can fall apart quickly if they do not understand the legal system on which their plans depend. Take this case from West Texas.

The West Texas sun shines bright and hot. Many have looked to the sky as a new form of investment. A rancher invested nearly $50,000 in solar panels to help make his operation more efficient and profitable. Later, the company

1. Special thanks to Texas Tech University student assistants Leigh Wells, Xan Argo, Barrett Moreland, and Rudy Benavides.

returned to fix a few issues, but this work was not done correctly. When the rancher challenged the faulty work, the company ignored him. So, as most people would, he considered taking them to court.

When the company installed the panels, standard paperwork was signed and completed by the rancher. Buried in the paperwork was an arbitration clause stating that he was not allowed to sue the company. If he had a grievance, he would have to go to a specific arbitration center, and an act as simple as filing the complaint could cost him $2,500. All this before even getting a hearing to discuss the problem.

At this point, the rancher was stuck. The repairs would have cost less than arbitration, and the company was aware of this. If the rancher had understood the system better (or read the contract fully), he could have pushed for a different agreement or walked away from the deal entirely.

This type of challenge happens more often than one might think, and not just with solar panels. It also applies to many land leases, equipment loans, inheritance plans, and a plethora of other documents and agreements. When people are unaware of what they are signing or of what their options are, they lose or weaken their own power.

That's where mediation comes in. Unlike arbitration, mediation is voluntary and affordable and allows both sides an opportunity to provide their perspective. It is not about handing power and authority over to an outsider. It is about helping families discuss issues before they escalate or get dragged into costly legal battles.

> *Question: Have you signed documents or agreed to terms that you have not read? How could this impact your estate or the future of your property?*

UNDERSTANDING CONFLICT

Conflict can develop from something small, such as a minor disagreement, or something more critical, such as an argument over a status or title. Conflict is a struggle arising between people with different needs, values, or goals. As Johnson and Johnson (2005) note, there are some central ideas to keep in mind when conflict occurs: You should face the conflict, reach an agreement that both parties can agree upon, and maintain appropriate relationships between the impacted parties. In the case of conflicts that might occur in family or land succession planning, all these points are incredibly important to consider. Lencioni (2002, p. 202) identifies "fear of conflict," or failing to recognize the power of "productive conflict in order to grow," as one of the most common pitfalls. People are hesitant to engage in conflict because of the discomfort it typically causes and because they do not realize the potential it offers for growth and development (Feirsen and Weitzman 2021; Headrick and Young 2023).

TYPES OF CONFLICT

To help us further identify the type of conflict that might be occurring, it is important to understand the types of conflict we can encounter.

- **Interpersonal** (between people). Examples: siblings, blended families, neighboring properties, extended family.
- **Intrapersonal** (within oneself). Examples: jealousy, guilt, falling out of love, personal injury and the emotional toll, mental health.
- **Structural** (caused by systems, roles, expectations). Examples: obligations to federal/state levels (documents, taxes, etc.), liens on property, ranch/farm management, child support, windmill/oil/mineral rights and leases, LLC or corporation status, power of attorney.

Once we have identified the type of conflict we face, we can begin to assess the situation and chart the path we take toward conflict management. In an agricultural context, conflict can originate from a variety of places. As succession and inheritance planning are discussed, for instance, uncertainty in how to move forward or what steps to take can cause conflict. When decision-making power or leadership roles are questioned or challenged, particularly in estate planning, conflict can occur. The practices and upkeep of a ranch or farm can depend on how the land is used and how environmental values are enacted; hence, conflict can occur. Another source of conflict might arise when there are financial pressures or generational differences in the way things are assumed to be managed. A further source of conflict might result from cultural or familial norms around the ways people communicate and share emotions and values. The phrase "We don't talk about feelings here" might have opened the door to conflict. Conflict allows us to view how we handle value-driven conversations and family expectations from a different vantage point. However, change is inevitable, and change can be difficult for people to process.

SOURCES OF CONFLICT

We have discussed the types of conflict, but what are the sources? According to Katz (1965), these sources generally revolve around economics, value, and power. **Economic conflict** involves competing motives to attain resources. You might be familiar with the term "my piece of the pie." Apportioning resources can lead to disagreement over who gets the largest "piece of pie." A second source of conflict, which centers around disagreement in personal interpretation of beliefs, values, morals, or family legacy, is called **values conflict**. Our personal interpretation of these beliefs might be similar but not always in step with those of our family members. The last source of contention is **power conflict**, which focuses on the maintenance or maximization of influence in decision-making or day-to-day management. At the heart of conflict is the need to control something. Pertinently, when one party is advantaged, another party experiences some sort of disadvantage. In metaphorical terms, their piece of the pie is too small; they don't even have a fork with which to eat their pie; and so on.

Conflict can manifest in several ways. There can be a silent tension around a topic of discord. Others display passive-aggressive behaviors, meaning they will be vocal with negative feelings to a certain point and typically will not face the conflict head-on. Some may avoid disputative people or difficult conversations altogether. Abrupt bouts of rage or emotion might be causing blow-ups about things not related to the actual conflict. Some individuals withdraw from situations outright if they fear a challenge. How we deal with conflict is heavily entwined with how we are taught to take it on, but the above examples provide some signs that conflict is brewing.

UNDERSTANDING THE ROLE OF CONFLICT

Conflict, while often viewed negatively, can in fact have positive outcomes. Thus, it is crucial to take on the difficult work of understanding the role of conflict in the situation you are experiencing. When we understand this point, as well as the type of conflict outlined above, we can be better positioned to address the conflict and move forward for the *best outcome available*. Notice that we used the term *available*, as sometimes the *best outcome possible* is not really possible given external factors like time, resources, people, and a variety of other things we do not have control over. Sometimes reaching the best outcome available allows us to make other decisions related to long-term planning and ways of navigating future conflicts around the best outcome possible.

We realize that conflict can be relational and situational in nature. This means we can recognize the difference in the way we view an issue and the

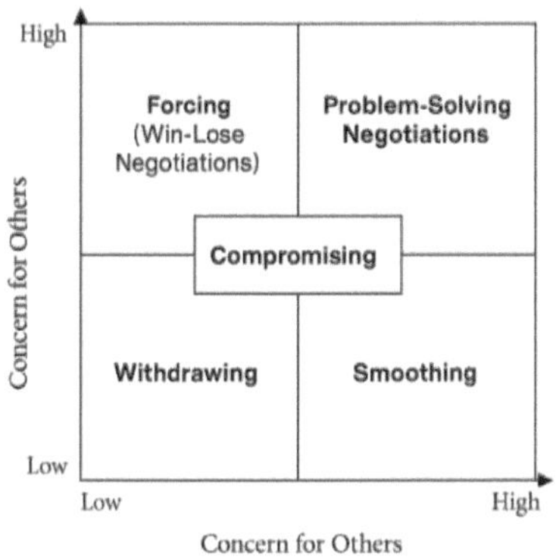

Figure 4.1. Five conflict management strategies. (Adapted from David W. Johnson and Frank P. Johnson, *Joining Together: Group Theory and Group Skills*, 12th ed. Pearson, 2017, 379–80.)

way others might relate to the same issue; this also means that the situations we face can absolutely factor in to how we deal with the conflicts and our level of engagement. Regardless of the view or the situation, conflict can provide a space for transformation and growth, even when one is opposed to the idea of engaging in difficult discussions with others.

HOW TO DEAL WITH CONFLICT

Both authors of this chapter are frequently asked for guidance on navigating conflict in daily life, drawing from their distinct professional backgrounds. While Dr. Jason Headrick primarily works with industry leaders and individuals on leadership and group dynamics, Attorney Gene Valentini manages the legal and procedural aspects of dispute resolution. Together, they provide a comprehensive approach to managing disagreements, beginning with five foundational models used to navigate and resolve conflict. Determining which strategy is appropriate can require you to consider your relationship with those with whom you are in conflict and what the goals are for the conflict to be addressed.

FIVE CONFLICT MANAGEMENT STRATEGIES

While these five strategies are often used in business and educational settings, they allow you to navigate the conflict by having options for your actions and next steps. Fig. 4.1 illustrates the strategies, with each defined in list form below.

1. **Problem-solving negotiations**. In this strategy, all members involved confront the conflict and identify ways to solve the problem together. This approach can help all members achieve their goals of resolution and relieves tensions and negative feelings. This strategy requires time and commitment to the overall goals and values associated with the conflict. It requires high levels of trust and a commitment to maintaining the relationships involved long-term. This can be suitable for family conflicts but does not work well if the decision to be made is urgent or if there is a loss of respect or communication for intentional discussions to take place.
2. **Smoothing**. This strategy can be considered when relationships are more important than the goal of complete resolution, as this approach requires self-sacrifice. This method can be used when the other parties' perspective and interest in the conflict are perceived to supersede yours, or if you discover you are wrong about the issue. If you are both equally invested in the outcomes, this is not the appropriate approach.
3. **Forcing or win/lose negotiations**. This approach can be used when future relationships are considered to be of little importance. Forcing can be viewed as a power move and elevates your concerns at the expense of others. Forcing tactics are not always the most civil, as they might include aggression or penalties of some kind. These approaches involve coercion, arguments, hard deadlines, and making demands that exceed acceptability. This tack will not win you any awards for relationship-building or creating a cooperative environment.
4. **Compromising**. This approach allows each party to meet in the middle and works well when both the outcome and the relationships are important. Each party gives something up in this approach. This does not always work well with creating long-term outcomes for complex issues. You both will suffer from some sort of loss, but the long-term result is best for others.
5. **Withdrawing**. This approach is where you or someone else avoids conflict and refuses to engage with it in any way. This can happen when others feel they do not have the power or necessary information or are too emotionally invested to engage in

discussion. This can cause increased tension and avoidance of other necessary decisions that have to be made. This approach can have lasting adverse consequences.

If you are someone who finds themselves navigating conflict and looking for some helpful hints on how to lead in these instances, it is important to know and understand that it is not "you against the world." When possible, you can share responsibilities and roles with your family members or others with whom you are in conflict. This means you may each be assigned to gather information, make calls, or research and report back to all parties involved. When you share the responsibilities, you are acknowledging that you trust others to be able to find the information, which can help ease the atmosphere and emotions that might be connected to the conflict. This trust can also play a big part in creating a psychologically safe environment. The idea of psychological safety basically affirms that people feel free to state their opinion or to share their feelings and emotions with others (Obrenovic et al. 2020).

CASE STUDY: FROM CHORES TO CRISIS—GIVING CREDENCE TO SAFETY

On a quiet family farm nestled in rural Pennsylvania, a middle-aged Amish farmer hitched up a wagon for a routine tobacco haul. He placed his two youngest kids—ages three and four—on the small horse-drawn cart that led the way. There was no railing on the wagon, and he intended this to be a short ride like the ones he remembered from his own childhood.

But this time, his daughter fell off the wagon. Tragically, the wheel of the wagon crushed her before he could get to her. She died instantly.

The incident triggered a criminal investigation, and local authorities reviewed the conditions of the wagon, the children's ages, and the lack of supervision. Prosecutors filed a felony child endangerment charge, citing state guidelines that recommend children under seven not engage in or be near farm equipment. The father did not contest the facts. A bench trial was held before a judge, who found the man guilty. He was sentenced to up to twenty-three months of confinement—most of which had already been served in pretrial detention—followed by five years of probation and fifty hours of community service focused on agricultural safety education.

But the legal sentence was nothing compared to the silence that filled the home after. The father's surviving children would not return to the fields that fall. His wife stopped going outside. By winter, the man was thinking about selling everything. What once felt like tradition now felt like a trap.

Incidents like this do more than just break hearts; they break succession plans. When a tragedy such as this hits, families often step away from the land entirely—whether out of grief, fear, or shame. A farm or ranch that was meant to stay in the family may instead be sold off to avoid the hurt and harder conversations that might follow.

Modern ranches and farming operations are not always the simple way of life they once were. In the modern age, these operations involve ATVs, augers, tractors, and industrial equipment. One lapse in supervision can mean a permanent loss.

Meanwhile, young people still need to be involved. Learning to work the land remains the best way to inherit an agricultural operation. These are reasons why tractor safety courses and ATV certification programs are essential for children, even when they were not considered a necessity in generations prior. Many ranchers write safety expectations directly into their succession plans. It's not just about avoiding accidents but also about making sure the next generation feels like their safety is prioritized and that they have been allowed to learn in their own ways.

> *Questions: Have you ever let a child help with something on the ranch that, in hindsight, might have been too dangerous? If something went wrong tomorrow, would your farm have the insurance, documentation, and training history to protect the family?*

THE ROLE OF MEDIATION

What do you do if a conflict is too big for you to take on through just conversations that seem to go nowhere and always end in the same insults and misguided actions? Who do you turn to if you need advice? It might be time to consult legal services and to consider mediation. Mediation is one problem-solving intervention that has proven to be effective in a variety of disputes and conflicts. It does not focus primarily on past behavior and specific weaknesses or emotional problems of individuals. Instead, it is future-oriented and builds upon the strengths of the parties involved to work with each other.

While there are numerous definitions of mediation, Texas law defines the term as "a forum in which an impartial person, the mediator, facilitates communication between parties to promote reconciliation, settlement, or understanding among them. A mediator may not impose his own judgment on the issues for that of the parties" (Texas Civil Practice and Remedies Code).

Basically, mediation is when two parties who have a problem with each other for one reason or another sit down and have a conversation with a person trained to help them resolve their differences. Sometimes this leads to a formal agreement like, "I'll keep my tree limbs trimmed if you stay off my property." And sometimes an understanding is reached, like, "I see that I was wrong for coming onto your property to trim limbs off your tree."

Consequently, the goals of mediation are broad:

a. Help parties to define their dispute and conflict in a confidential setting
b. Empower the parties to resolve their differences and deescalate their controversy
c. Allow individuals an opportunity to express their feelings
d. Enable parties to restore a relationship if desired.

These goals often contradict how individuals are socialized about disputes and conflicts. The positive aspects of conflicts may be ignored because of the emotional desire to win at any cost. Communications are stopped and thus information can be misinterpreted, which then possibly expands the harsh feelings between the involved parties. No or limited communications lead to the most important role of a mediator: effective facilitation. A mediator is not the decision-maker for resolving the dispute; that task belongs to the involved individuals. So, what can be mediated?

TYPES OF MEDIATION

Family disputes often benefit from using the mediation process to resolve a range of topics such as:

- Transitioning of ranch and/or farm
- Financial issues within family or with external lenders
- Succession planning
- Divorce / parenting
- Communication / family dynamics
- Management concerns
- Lifestyle changes / challenges

Any controversy that is challenging for individuals to discuss and resolve is probably something that can be handled by an effective mediator.

PRE-MEDIATION

If a controversy or dispute develops, holding a pre-mediation discussion with a mediator might be constructive. The mediator can meet with the invited parties prior to commencing mediation and go over the desired topics for the actual mediation. This enables the attendees to begin thinking about resolutions that could resolve the dispute. After a mediation agenda is developed, individuals can begin developing possible solutions that can be discussed at mediation.

USDA-FUNDED PROGRAMS

Nearly every state has a USDA-funded Certified Mediation Program that can be utilized for farm and ranch family disputes. The respective state governors designate an entity to provide the service and receive federal funding. For over two decades, Lubbock County's Texas Dispute Resolution System has provided mediation for disputes free of charge. Mediation can be in person, via telephone, or virtual. For family-related issues, the preference is for an in-person meeting. The Texas Department of Agriculture (TDA) can also be involved in deciding what type of issues can be mediated. Federal law governs the funding and parameters for administrative matters.

CASE STUDY: BROTHERS WILL BE BROTHERS

In rural Iowa, an aging farm couple owned 980 acres, including land that had been inherited and some that had been purchased acre by acre over the years. Their three sons—Earl, Wayne, and Dale—had each taken a different path, but all had a connection to the land. Earl stayed on the family farm. From the time he was a teenager, he worked full-time beside his father, eventually co-owning eighty acres. Wayne lived nearby and helped seasonally during planting and harvest, though he had a full-time job in town. Dale lived farther away; having earned a degree in finance, he handled the farm's books.

As the parents grew older, the question of what would happen to the farm became harder to ignore. Each brother had a different expectation—some spoken, some not. Wayne and Dale wanted to discuss the future, but Earl was protective of his contribution and reluctant to debate the issue. Wayne felt he'd contributed more than Earl gave him credit for. Dale worried he'd be left out completely. Earl, having sweated the daily grind, believed he'd earned more than just a third of the holdings. The parents, caught between their sons and afraid to stir resentment, said little and the planning stalled.

Eventually, tension reached a point where the family agreed to sit down with a mediator from the Iowa Mediation Service. At first, it wasn't easy. There were years of unspoken assumptions and issues to unpack. But the mediator helped each brother articulate what he wanted—and more importantly, what he feared. Earl shared his anxiety about losing control. Wayne talked about being overlooked. Dale asked for transparency and fairness.

In the end, they crafted a plan that reflected both the history and the future of the farm. Earl would continue running day-to-day operations. Wayne and Dale would receive compensation proportionate to their involvement. More importantly, the family set up regular meetings to talk about the farm, not just run it. The friction did not disappear overnight, but at least an environment had been created where it could be approached and discussed. Through structure, respect, and a little outside help, the family found its way back to the same table.

> *Questions: What assumptions are going unspoken in your family? Has each person's role on the ranching / farming operation been clearly acknowledged? Think about a conversation you've been avoiding regarding succession—will it get easier to talk about the longer you wait?*

KEEPING CONFLICT IN CHECK

Ranches, farms, and families can take action to keep conflict in check. Regular meetings of everyone involved, from workers to owners, can help to prevent emotional bottlenecking in the operation. You can call these gatherings family councils or operational check-ins. They can constitute structured decision-making processes and can further allow for generational mentoring. We have all heard members of the different generations object that others have not involved them in discussions on family affairs, and their knowledge is often gleaned from overhearing information or stumbling upon it later. Consistent meetings strengthen family connections and communication.

SELF-ASSESSMENT QUESTIONS

1. Who are the parties involved?
    a. If only family, what is our history of resolving conflict as a family?

b. If others are involved, what is our history of conflict between our operation and others?

2. What is the source of conflict?
3. What type of conflict (interpersonal, intrapersonal, or structural) are we experiencing?
4. How do we see the conflict happening?
5. We typically understand our perspective of conflict. What is the other party's perspective (not your personal diagnosis, but thoughts and feelings) of the conflict you are experiencing?
6. How have you tried to resolve the conflict?

As regards mediation:

7. Does everyone in the family agree with the proposed succession plan?
8. Do any sibling issues need to be addressed?
9. Do any blended family issues need to be addressed?
10. Do any peer issues need to be addressed?
11. Have any promises been made that should be discussed?
12. Are there any capacity or health issues that could impact everyone?

CHAPTER 5

Shaping Milestones, Road Maps, and Family Meetings

Amita Quenim and Anish Quenim

INTRODUCTION

As the chapters come to a close, we can conclude that succession planning for ranches is a complex and emotional journey that requires planning, learning, and implementation. Unlike a corporate handoff, the passing down of a ranch often represents generations of sweat equity, memories, identity, legacy, and livelihood. The success of this transition depends not just on memorandums and handshakes of understanding but on clear milestones, intentional road maps, and meaningful family conversations to corroborate the processes. This chapter aims to provide a practical guide to succession planning with a focus on understanding milestones, developing a succession road map, leveraging family meetings, and taking action-oriented next steps. Whether you're an individual rancher, a couple managing a ranch, a multigenerational family working together, or from a ranching family, this chapter seeks to help you navigate the path to a successful and sustainable business, family, and family business succession.

As part of an exclusive for this chapter, we had the pleasure of interviewing John Welch and his son, Wesley. Both father and son spoke about the generational transition of the family-owned Welch Cattle Company, and where they are today, both as a family and a family business.

SECTION 1: UNDERSTANDING MILESTONES IN SUCCESSION PLANNING

Imagine driving without mile markers. The route seems endless, and the next turn is unclear. Similarly, milestones are critical markers in the succession

Figure 5.1. A gathering of the Welch family, representing the multigenerational continuity of the Welch Cattle Company in West Texas. This image illustrates the intersection of family heritage and business operations, highlighting how the transmission of traditional ranching values and shared identity supports the long-term succession and economic resilience of the enterprise. (Courtesy of the Welch family.)

planning process. They help families track progress, make informed decisions, and ensure accountability at every turn. For the context of the families in ranching succession toolkit, milestones can be categorized into five primary areas, each addressing a key aspect of preparing the ranch and the family for transition.

•

Psychological Insight: Navigating the Emotions Behind Succession

Before you dive into the milestones described below, it is important to acknowledge what often goes unspoken as emotional undercurrents that make ranch succession uniquely challenging.

For the generation in place, the thought of letting go can trigger *anticipatory anxiety*, a fear of the unknown, of losing purpose, or of what life will look like outside the daily operations. The ranch is not just a business; it's an identity, a legacy, and often comprehends a lifetime of work. The transition can feel like surrendering a part of oneself.

For the incoming generation of successors, *change resistance* can surface as the weight of expectation grows. Taking over a ranch comes with the pressure to prove one's worth, honor family traditions, and manage both people and property effectively. The fear of failure, or of not being fully trusted, can be challenging, daunting, and haunting.

This section helps families move forward not just strategically but also emotionally. The milestones outlined ahead provide structure, clarity, and a sense of shared progress. By breaking the process into manageable parts, families can reduce uncertainty, build confidence, and ease the emotional load that often accompanies big transitions.

•

1. **Personal Readiness Milestones**. Personal readiness involves self-reflection and preparation by the generation in business. This includes gaining clarity on personal values and a vision for retirement or reduced involvement. It also requires emotional readiness to step back from daily operations and consideration of health and lifestyle needs in planning for a new phase of life, for all connected stakeholders.

TAKEAWAYS

- Succession starts with the current leader's clarity and willingness to transition.
- Emotional and physical readiness significantly impacts the success of the process.
- Personal goals should align with the future of the ranch and family values.

ACTION ITEMS

- Write down your vision for life after succession, including what the change in lifestyle could look like, and continued involvement levels, if any, with the family ranch.
- Create a personal wellness plan for the transition period. Wellness programs will vary from person to person but ensure personal productivity in the many more hours that will soon be available.

- Reflect on your emotional attachment to daily operations and discuss these openly with a trusted advisor and the incoming generation. This effort often translates into legacy continuity.

2. **Successor Development Milestones**. Preparing for the next generation is vital for a smooth transition. This includes identifying a potential successor (or successors) and supporting their development through education and hands-on experience. As they grow, successors should be given increasing responsibility and authority in managing ranch operations to build confidence and competence. Refer to the 30 SPLIT Succession tool, developed by the Quenim family, featured toward the end of this section.

TAKEAWAYS

- Identifying and grooming the right successor is an intentional, long-term process.
- Successors need both formal education and hands-on experience to lead effectively.
- Responsibility should increase gradually to build confidence and capability.

ACTION ITEMS

- Create a successor development plan with timelines, education goals, and milestones.
- Assign your successor a new operational responsibility and analyze progress. Consider intrapreneurship as an option—i.e., the new entrepreneur within the family business.
- Arrange mentoring or shadowing opportunities with other successful ranch operators.

3. **Business Readiness Milestones**. A ranch must be operationally sound to support a successful transition. Business readiness includes maintaining accurate financial statements, ensuring transparency, and having a clear understanding of the business model and profitability. Formalizing operations with

documented policies and procedures also strengthens long-term sustainability and supports a smoother handoff. These operating procedures, manuals, and files must be made accessible to the successors?

TAKEAWAYS

- Transparent and organized operations are essential for building successor confidence.
- Understanding financial and legal health enables sound decision-making.
- Documented procedures ensure business continuity across generations.

ACTION ITEMS

- Hire a professional to review and organize your financial and business model.
- Develop or update your ranch's operations manual, including operating procedures.
- Hold review meetings with successors and provide them with access to the archives. Design a trend versus actuals projection for the future. This helps with pre-planning and forecasting. Be mindful to consider the needs of both the growing business and the family.

4. **Legal and Estate Planning Milestones**. Legal and estate planning is essential for protecting assets and ensuring clarity. Key milestones include updating wills and trusts, establishing power of attorney and healthcare directives, and aligning buy-sell agreements and business structure with succession goals. These legal steps reduce uncertainty and minimize potential disputes during the transition. For more information on this topic, refer to chapter 3.

TAKEAWAYS

- Without legal alignment, even the best succession plans can unravel.

- Estate planning tools like trusts and wills must be kept current.
- Clear agreements prevent future conflict and confusion.

ACTION ITEMS

- Meet with an estate attorney to review and update all legal documents.
- Draft or revise family agreements that align with succession goals.
- Organize all legal paperwork in a secure, accessible location and inform family members of its existence. Remember to invite participation and input from family at the onset.

5. **Relationship and Communication Milestones**. A family that encourages open lines of communication, establishes communication boundaries for business and family discussions, and provides a platform for each member often succeeds. Healthy family dynamics are foundational to succession success. Milestones in this area include holding regular, structured family meetings, establishing conflict resolution mechanisms, reaching agreement on shared values and long-term vision, and knowing when family time and business time demands and deserves separation. Open and consistent communication fosters trust, reduces misunderstandings, and helps keep everyone aligned throughout the process.

TAKEAWAYS

- Strong relationships support smoother transitions and sustained trust.
- Communication needs to be structured and allow room for informal family time.
- Alignment on vision and values strengthens family unity. Let this be known as "the legacy narrative"—the shared story that defines who the family is and where the ranch is headed. Action Items
- If you haven't already, start holding regular family meetings with clear agendas and roles.

Figure 5.2. A member of the Welch family demonstrates traditional roping techniques during calf working. (Courtesy of the Welch family.)

- Establish a written family vision statement and revisit it annually. Be mindful of age groups and foster plans that encourage the family's participation for the success of all involved.
- Develop a proactive, reactive, and rehabilitative conflict resolution plan that includes outside facilitation if necessary. To support this structure, ensure pathways to communication are established, and safe, honest, and protected spaces are created.

CODA

Each milestone contributes to a comprehensive and realistic succession plan. Families should take time to evaluate their status in each of these five areas and identify any gaps. By addressing these milestones methodically, ranch families can build a strong foundation for generational continuity and long-term success. We're here to help you as well!

REAL VOICES: THE WELCH FAMILY'S SUCCESSION JOURNEY

An example of the milestone principles in action is the story of John (father) and Wesley (son) Welch from the Welch Cattle Company.

John Welch explains:

> *First came the decision to leave Texas and start over in Colorado. We [my wife and I] gave up our careers as a couple to go all in on building the family business. From there, it was about taking those first bold steps to get it off the ground. We involved our sons from an early age, not just to help with the business but to raise them in it. More than anything, we wanted to give our children the tools to chase their dreams, to grow, and hopefully, one day, to come back and make this legacy their own. I didn't have a big legacy handed to me; I had to create it. And if the day ever comes when decisions must be made about the business, even if that means selling, it shouldn't come with pressure to keep it just for tradition's sake. For us, preserving the land isn't just about running a business. It's about preserving what holds the family together.*

Wesley Welch observes:

> *The best thing about this journey has been a lifestyle learned from childhood, through upbringing, and deeply preserved and practiced by the siblings. The transition from son to peer with my dad took time, but it did happen, with effort. Now I play the same role with my son.*

The Welches' story mirrors the heart of succession, rooted in values, evolving through action, and always moving toward continuity—a process built on steps taken, one step at a time.

THE 30 SPLIT SUCCESSION RULE

Succession is more than a milestone; it's a rhythm. After reviewing the five milestone categories listed above, consider applying the 30 Split Succession Rule as a practical, generational model to guide your long-term transition planning.

The 30 SPLIT (Succession Processes Learnings in Transition) model, developed and successfully employed by Amita Quenim (mother) and Anish Quenim (son), reimagines succession as a 30-year, three-phase cycle that encourages long-term stewardship and sustainable leadership development. It helps families see succession not as a finish line but as an evolving journey of growth, trust, and preparation.

Side note: Keep in mind that succession is an ongoing process, from pre-birth until post-death, and the continuing need to honor the legacies of the departed. Although we recommend three potential phases, elaborated over 30 years, the efforts you put into these family relationships before and after set the precedent for a successful succession process. Start early and never stop. As mother (and new grandmother) and son (and new father), we see it too!

The Three Phases of the 30 SPLIT

1. **Apprentice Phase** (years 1–10): Learn from the outgoing generation.

    - Successor works closely with the current leader.
    - Activities: job shadowing, operations support, mentorship, and values immersion.
    - *Goal:* Develop a deep understanding and build foundational competence.

2. **Leadership Phase** (years 11–20): Independently lead the business.

    - Successor fully steps into leadership.
    - The outgoing generation steps back, advises through an advisory board, and/or moves over to managing other non-core avenues under the family umbrella.
    - *Goal:* Execute vision, lead with confidence, and grow strategically.

3. **Mentorship Phase** (years 21–30): Develop the next successor.
    - The current leader becomes a mentor to the future generation.
    - Begin identifying, nurturing, and transferring knowledge to new leaders.
    - *Goal:* Ensure wisdom transfer, relational stability, and long-term continuity.

Why This REPEAT CYCLE Works

- **Continuity**: Ensures no generation exits without passing down knowledge.
- **Clarity**: Each phase has a purpose, preventing rushed or vague transitions.
- **Confidence**: Both outgoing and incoming leaders have time and space to grow.

SECTION 2: CREATING THE SUCCESSION ROAD MAP

Continuing the driving reference: From manual maps to digital maps, establishing the destination and following a navigation guide takes us from point A to point B. Similarly, a succession road map serves as a customized guide to help families transition ranch ownership, management, and leadership smoothly. It translates complex, long-term goals into achievable and actionable phases, providing clarity and structure to a deep personal process. With a clear road map, families can navigate change with confidence and reduce uncertainty for everyone involved.

•

Psychological Insight: Understanding Roles in the Family Business System

Before diving into succession road maps, it's essential to recognize that families are not just collections of individuals: They are systems, i.e., many moving parts coming together to form the whole. According to Systems theory, each member unconsciously plays a role, in this case to help unite the business family for success. These roles influence how members behave, communicate, and respond to change. Be mindful and humble to acknowledge family members often unseen or not celebrated enough, for their subtle contributions are consistently impactful.

When succession planning begins, these ingrained roles come to reflect. The founder who has always led must consider stepping back. A child long perceived as "not ready" may be stepping into authority. Spouses, siblings, cousins, and in-laws may have perspectives shaped more by family dynamics than business logic. If left unaddressed, these roles can create conflict, resistance, or confusion about what's fair, expected, or possible. Let's call this "family governance."

How individuals see themselves in the context of family and business (role identity) also plays a significant part in family dynamics. One may be a rancher, a daughter, a leader, and a caretaker all at once. These identities intersect, and when transitions occur, the relevant party can feel threatened or confused. That's why succession road maps must be both individual and collective, allowing space for personal growth, relational negotiation, and family alignment.

As you explore this section, remember: The road map is not just logistical. It is emotional, psychological, and relational. It's a blueprint for navigating identity shifts, balancing power, and evolving together, so that the ranch's legacy includes not only land and operations but also healthy, resilient relationships. Understands the hats each one wears, for overall success.

•

Step 1: **Define the Vision**. Clarifying the future vision is the foundation of succession planning. Families should ask: What does success look like in five, ten, or twenty years? Who will own and manage the ranch? What legacy do we want to preserve? What role do we all play in celebrating our generation and raising the next? These questions ground the road map in shared purpose and long-term aspirations.

TAKEAWAYS

- A shared vision provides direction and unity throughout the transition.
- Vision planning must address both ownership and operational leadership.
- Legacy is more than land; it includes values, relationships, and community impact.

ACTION ITEMS

- Hold a family visioning session and ask each member to describe generational success.
- Create a written vision statement that outlines desired legacy, ownership, and leadership.
- Identify areas of alignment and disagreement to begin building consensus.

Step 2: **Conduct a Readiness Assessment**. Before progressing, it's critical to assess readiness across multiple dimensions: personal, financial, legal, and operational. Tools like SWOT analysis (Strengths, Weaknesses, Opportunities, and Threats) help families identify areas that need attention. Honest assessments allow for a realistic starting point and prioritize efforts where they're most needed. Let's bring out the whiteboards!

TAKEAWAYS

- Assessing readiness provides a reality check before initiating transition steps.
- Readiness spans personal, business, legal, and emotional factors.
- Gaps identified early can be proactively addressed.

ACTION ITEMS

- Use a checklist or self-assessment tool to evaluate key readiness areas.
- Conduct a SWOT analysis of the current ranch operation. Consider visiting with a FIRST professional for this.
- Meet with a trusted, knowledgeable advisor to review findings and prioritize action items.

Step 3: **Clarify Roles and Responsibilities**. Successful transitions depend on clearly defined roles: family governance. Map out responsibilities for both current and next-generation leaders and the roles for the family in support positions—remember the unseen, less celebrated ones? Consider transitional options like co-leadership, mentorship roles, or phased leadership handoffs. This clarity helps avoid misunderstandings and sets expectations for decision-making, accountability, and succession.

TAKEAWAYS

- Clarity prevents conflict and confusion as roles evolve.
- Roles should reflect each person's strengths and developmental needs, for the family and business.
- Transitioning roles gradually can support smoother leadership shifts.

ACTION ITEMS

- Create a chart of current and future roles for each family member involved.
- Outline key decision-making responsibilities and assign authority levels.
- Explore co-leadership or mentorship structures during the transition phase.

Step 4: **Set Goals, Timelines, and Contingency Plans**. Turn the vision into tangible results by setting SMART goals (Specific, Measurable, Achievable, Relevant, and Time-bound). Break long-term objectives into smaller, manageable actions with clear deadlines. A structured timeline keeps the family on track and allows for monitoring progress along the way, reducing delays or indecision. Even the best-laid plans need backup. Prepare for unexpected illness, death, or a change in interest by creating contingency measures. Identify alternative successors and establish emergency protocols. This plan ensures the ranch can continue to operate effectively under pressure or sudden change.

TAKEAWAYS

- Concrete goals turn intentions into measurable progress.
- Timelines promote accountability and momentum.
- Breaking big goals into smaller steps avoids clutter.

ACTION ITEMS

- Develop SMART goals related to succession within the next year, for all ages.
- Assign tasks and target dates to each goal in a shared calendar or planning document.
- Review and revise goals at family meetings based on progress and new developments.

Step 5: **Document the Road Map**. Put everything in writing and provide accessibility to all. You may use visual tools like charts, timelines, and checklists to make the road map easy to follow. Documenting the plan ensures continuity, especially during leadership transitions, and makes it easier to communicate expectations. Review and update the road map annually to keep it relevant and responsive.

TAKEAWAYS

- A written road map makes expectations visible and accessible.
- Visual tools enhance clarity and engagement across generations.
- Regular updates ensure the road map remains relevant as conditions change.

ACTION ITEMS

- Create a one-page summary of your road map, including steps, goals, and timelines.
- Schedule an annual review of the road map with all stakeholders.
- Include the element of flexibility. Life happens to us all, and pivoting is key.

CODA

A well-documented road map ensures that everyone involved understands what's coming next and what is expected of them. It reduces confusion, fosters accountability, and keeps the family aligned through every phase of the transition, deeply rooted in a formal family governance structure. With a clear path forward, families can steward the ranch and a generational legacy.

SECTION 3: THE ROLE OF FAMILY MEETINGS IN SUCCESSION PLANNING

Family meetings are the cornerstone of open communication in the succession planning process. They provide a structured environment where family members can share perspectives, make decisions together, and address sensitive business and family issues. These meetings help ensure that everyone feels heard and that the planning process stays transparent and collaborative. Family meetings can be formal or informal. While the majority of this chapter focuses on formal meetings, we will conclude by addressing the importance of informal family meetings as well.

•

Psychological Insight: Why Family Meetings Aren't Just About Logistics but Serve as Emotional Repair Rooms

Before reading this section, take a moment to consider this: Succession conversations often awaken old wounds. Family meetings, while intended to be productive, can stir up unresolved tensions, hurt feelings, or power struggles, especially when discussions around leadership, fairness, or future roles come into play.

The absence of communication is rarely neutral; it often breeds assumptions, unless discussed, deliberated, and resolved. And assumptions, left unchecked, can fester into resentment. When families gather to talk about the future, it's not uncommon for unspoken histories to reappear, sometimes subtly, sometimes explosively. The key is to meet. We have often noticed that all it takes is a cup of coffee, some favorite cookies, and a comfortable room to foster progression.

That's why active listening and validation are foundational to successful family meetings. These aren't just tools for "good behavior," they are healing mechanisms. When a family member feels truly heard, not interrupted, not dismissed, their defensiveness decreases, and their willingness to collaborate increases. First listen and seek to understand, then contribute to the talk.

Figure 5.3. The extended Welch family gathered outdoors on their ranch property. This wide-angle view emphasizes the size of the family unit and their collective presence within the rural West Texas landscape they steward. (Courtesy of the Welch family.)

This chapter is about more than planning meetings. It's about practicing emotional intelligence in a business setting. It encourages families to build a culture of listening, de-escalation, and respectful disagreement. When family meetings become safe spaces for both logic and emotion, they can be the place where old tensions are resolved and new trust is built.

•

WHY FAMILY MEETINGS MATTER AND THE PROCESS THAT FOLLOWS

Family meetings build trust by creating transparency. They foster inclusion, especially for those not involved in daily operations, and provide a space for education about the ranch and its future. Regular communication helps reduce misunderstandings and prevents long-term conflict that could derail succession efforts or strain family relationships.

Wesley Welch shares:

We have an annual family meeting in March. My brother Andy talks about financials and runs the projections for next year. It is also a chance to bring cousins together. Families gather the night before, children play, we talk about personal things, and not business. The next day, we get down to business, talk till noon, have lunch, and go our separate ways. These times together allow us to reiterate our togetherness, and there have been times when I have gone through things, and these meetings provide that much-needed support and advice.

John Welch notes:

As a father, my sons and I would ride our horses and talk about business and life. But the March family meeting is purely focused on the business agenda. The three sons run the meeting, and my wife and I assist as needed. My wife Bonnie ensures that once a year, we take a family trip—and that trip is strictly holiday time, no business talk. She makes sure the family-time schedule is packed so that no one can bring up business. Families must learn this difference, and those lines of separation.

As David C. Bentall writes in his book *Leaving a Legacy*, "What separates thriving family enterprises from those that fracture is not just financial discipline, but emotional maturity and shared purpose, both of which are nurtured in the rhythm of regular family meetings."

1. **Set a Regular Schedule**. Consistency is key to making family meetings effective. Whether monthly, quarterly, or twice a year, setting a regular cadence helps normalize these conversations. Families should choose a frequency that aligns with their level of complexity and stage in the succession process, ensuring continuity and keeping momentum throughout planning. Most importantly, set the environment free from distractions and any family triggers—i.e., choose a safe and dedicated meeting space.

TIPS

- Use a shared calendar to schedule meetings in advance, and ensure calendars align.

- Tie meetings to natural business cycles, such as before tax season or after harvest.
- Stick to the schedule, even if there's no crisis; consistency builds habit and trust.

2. **Establish Ground Rules**. To foster respectful and productive discussions, families must set clear ground rules. These may include active listening with no interruptions and a commitment to confidentiality. Ground rules create a safe environment where members can speak openly without fear of judgment, fostering mutual respect and healthier conversations.

TIPS

- Ensure there is scope for both family and business discussions.
- Write ground rules down and read them aloud at the beginning of every meeting.
- Have everyone sign a "family charter" that includes communication expectations.
- Use a "talking stick" or speaking token to ensure only one person talks at a time.
- Encourage a "no surprises" rule; all big topics should be previewed before meetings. In short, stick to the agenda.

3. **Designate a Facilitator**. A designated facilitator helps keep meetings focused and fair. In some cases, an external facilitator, such as a consultant or mediator, can provide neutrality and guide discussions constructively; we have a recommendation in the chapter's conclusion. This is especially helpful when addressing emotionally charged topics or when the family lacks a natural moderator within its ranks.

TIPS

- Rotate facilitation among family members to build skills and share responsibility.
- For tough topics, hire a neutral third party, like a family business advisor or mediator.

- Brief the facilitator beforehand to review the agenda and agree on time limits.

4. **Use Agendas and Minutes.** Preparation enhances effectiveness. Every meeting should have an agenda distributed in advance, so everyone knows what to expect. Invite feedback on the agenda. Taking and sharing minutes afterward helps ensure accountability and follow-through. Documenting decisions, tasks, and open questions keeps the group aligned between meetings and builds momentum toward goals.

   TIPS

   - Send out agendas in advance so family members can prepare or add topics.
   - Stick to time limits for each item; use a timer if needed.
   - Assign a "scribe" to take minutes and send them out after the meeting.
   - Use shared documents or project management tools to track tasks.
   - Gentlemen's handshakes are a real thing, but agreements should ultimately be written down.

5. **Include Everyone Affected.** Succession affects more than just those running the ranch. Including spouses and non-operating family members in meetings, when appropriate, encourages broader understanding and reduces surprises later. These individuals often hold influence and insights that can enrich discussions and strengthen overall family unity.

   TIPS

   - Invite spouses or silent partners for high-level updates and buy-in.
   - Use small group breakouts during meetings to hear from the introvert voices.
   - Acknowledge different roles (family governance) to ensure that inclusion feels authentic.

TOOLKIT: TRY THIS FAMILY MEETING SIMULATION PRACTICE HOLDING A FORMAL FAMILY MEETING BY FOLLOWING THIS STRUCTURE:

1. Assign meeting roles: facilitator, participants, etc.
2. Set a date, time, and neutral location. Add to the calendars.
3. Choose a discussion question relevant to succession or operations.
4. Run the simulation for sixty minutes. Focus on listening and respectful exchange.
5. Debrief afterward: What worked well? What was difficult? Schedule a follow-up.

SIDEBAR: MAKE IT EDUCATIONAL

Family meetings are an opportunity to educate and empower the next generation. Invite FIRST professionals, or ranch consultants, to present on relevant topics. Share financial updates, operational insights, or lessons learned. Education builds capacity and helps successors feel more confident stepping into leadership roles.

TOPICS TO COVER IN FAMILY MEETINGS

1. **Leadership Succession Planning**. Discuss how future leaders will be identified, groomed, and evaluated. Talk about timing for transitions, shared responsibilities during handoffs, and how the outgoing generation can stay engaged meaningfully after stepping back. Refer to the 30 SPLIT model.
2. **Family Employment Policies**. Clarify who can join the business, under what circumstances, and what expectations are in place. Deliberate whether and how spouses or the next generation's children can be involved and how exits or underperformance are managed. Discuss remuneration, perks, growth, and other relevant topics.
3. **Career Development and Mentorship**. Set goals for the professional growth of family members in the business. Identify needed experiences, training, or mentorship. Make space to celebrate achievements and plan for future leadership roles. Start young and anticipate continuity, from school to graduation to business to eventual retirement.

4. **Compensation and Incentives**. Ensure there is transparency around how salaries, bonuses, and benefits are determined. Align incentives with the business's goals and ensure fairness between family and non-family employees. Nepotism is often a concern, unless transparent and structured fairly.
5. **Ownership Structure and Financial Expectations**. Review stock ownership policies, dividend decisions, and shareholder rights. Talk about how liquidity needs are addressed and ensure everyone understands their financial role and returns.
6. **Governance and Decision-Making.** Discuss the roles of the board, family council, or any other governance bodies. Decide how strategic decisions are made, who serves in leadership positions, and how the family participates in oversight. Decisions and discussions for family and business growth are mutually inclusive.
7. **Family Relationships and Conflict Resolution**. Create a forum for addressing interpersonal tensions and generational differences. Encourage open communication, empathy, and mechanisms for resolving conflicts constructively.
8. **Communication and Family Agreements**. Review and update family agreements, charters, or policies. Schedule future meetings, plan joint activities, and ensure that communication channels are strong, inclusive, and consistent.
9. **Values, Conduct, and Accountability**. Reinforce family values and expectations for personal behavior. Discuss how the family supports members during hardship and how it handles issues of misconduct or reputational risk.
10. **Community Impact and Legacy**. Explore ways the family can give back to the community and support charitable efforts. Recognize individual contributions, manage the family's public image, and promote entrepreneurial ideas from within the family. Often, families in business raise a village.

CODA: THE QUIET POWER OF INFORMAL FAMILY MEETINGS

While much of this chapter has focused on formal family meetings with structured agendas and business goals, the role of informal family gatherings cannot be overstated. These moments—shared meals, family trips, holidays, or even just coffee on the porch—are where the bonds of family are nurtured without the pressure of decisions. They are the "invisible threads" that hold a family together when succession conversations become tense or uncertain.

As John Welch noted, it is his wife Bonnie who ensures the family takes an annual trip with a clear boundary: No business talk allowed. She fills the itinerary with activities and shared experiences that allow every generation, children, cousins, and elders, to connect in laughter, stories, and joy. These times aren't a break from the business; they are an investment in the family that supports the business.

Informal gatherings should not be overlooked or left to chance. They deserve just as much intentionality, even if the tone is lighter. Having an informal "agenda," one that includes games, outdoor activities, storytelling, or even shared cooking, ensures everyone across all ages feels included and engaged. These rituals create shared memories, deepen trust, and build the emotional equity needed to weather future challenges.

In many ways, informal family meetings represent succession planning too, not through documents or decisions but through love, presence, and togetherness. A family that plays together, laughs together, and supports each other beyond the business is more likely to remain unified when it's time to make hard choices about the future.

A FINAL WORD ON SEEKING OUTSIDE SUPPORT

As families navigate the emotionally layered and strategically vital process of succession, even the most well-intentioned conversations can become complex. While many families are capable of hosting productive meetings on their own, there are times when a neutral presence, someone who understands both family dynamics and business realities, can make all the difference.

Having an experienced facilitator at the table can help keep discussions balanced, emotions in check, and decisions focused. Whether to guide a difficult conversation, include quieter voices, or simply create structure, a professional facilitator can help families preserve both their legacy and their relationships. It's not about outsourcing the family meeting; it's about supporting the family conversation with someone who understands the stakes.

SECTION 4: IMPLEMENTING THE NEXT STEPS

Once milestones are identified, the road map is created, and family meetings are underway, it's time to shift from planning to action. This is where succession becomes real. Implementation requires commitment and coordination and demands continuous communication from every member, whether leading, supporting, advising, or preparing to step in. Baby steps!

•

Psychological Insight: From Knowing to Doing—Overcoming Decision Apprehension

Succession planning often comes with a surprising twist: Even when the plan is clear, families can find themselves stuck in hesitation. This hesitation is rooted in what psychologists call decision apprehension, or the fear of making the "wrong" choice, hurting someone's feelings, or disrupting the family dynamic. Add in the complexity of overlapping relationships, responsibilities, and expectations, and decision fatigue becomes real.

Many families find comfort in planning, workshops, road maps, and family meetings, but then feel emotionally overwhelmed when it's time to act. For some, the fear of failure holds them back. For others, it's the emotional weight of change, or unresolved doubts about readiness, authority, or fairness. Even baby steps can feel disproportionately challenging.

This section is designed to help families move from knowledge to execution with clarity and confidence. It provides practical tools to assign responsibility, lock in decisions, monitor progress, and celebrate wins, while addressing the human need for reassurance, collaboration, and shared accountability. Whether you're an individual feeling uncertain, a couple navigating the future together, or an extended family with differing views, you don't have to be fearless; you just must be willing.

This is where your plan becomes practice, your values become action, and your legacy becomes lived. Saddle up, one step at a time, from "This might be my first rodeo" to "This ain't my first rodeo!"

•

1. **Delegate Tasks and Follow Through**. Assign responsibilities based on each person's strengths, interests, and availability. Don't expect one person to do everything. Use tools like checklists or shared calendars to stay organized. Schedule regular check-ins, monthly or quarterly, to monitor progress. Accountability ensures momentum, and shared responsibility reinforces a culture of cooperation and trust.
2. **Formalize Agreements**. It's essential to turn verbal commitments into written agreements. Legal documents, such as buy-sell agreements, succession plans, or ownership transfers,

should reflect decisions made by the family. Avoid informal promises or handshake deals. Clear, documented agreements protect relationships, prevent misunderstandings, and offer peace of mind to all generations involved.

3. **Invest in Education and Training**. Support successors with intentional development. Encourage them to attend FIRST workshops, engage in mentorships, or enroll in courses related to ranch management, finance, or governance. Families can also learn together, host educational sessions, bring in guest speakers, or join FIRST industry networks. Growing knowledge strengthens both confidence and continuity.
4. **Monitor and Adjust**. Succession plans are living documents. Schedule annual reviews to reflect on what's working and what needs to shift. Be open to adjusting the plan if there are changes in health, finances, or family roles. Flexibility and the ability to pivot do not constitute failurebut rather wisdom. Adaptation ensures long-term success without sacrificing the spirit of the plan. Keep hustling!
5. **Celebrate Milestones**. Don't forget to acknowledge progress. Celebrate the big moments, new roles, completed transitions, and anniversaries of succession steps. Host a family gathering, give symbolic gifts, or recognize contributions publicly. Celebrations reinforce shared values, honor hard work, and help the next generation feel proud and empowered to continue the legacy.

John Welch shares:

Succession was a continuous process that eventually evolved through baby steps with my sons, steps we took together. Steps include an evolving business and an evolving family. Now they're in charge, and the only decision I need to make is which horse I ride. I made sure the kids were involved each step of the way, and once they know how to run the business, bite your tongue, even though you feel the urge to advise.

Figure 5.4. Multiple generations working in tandem during a cattle sorting and roping exercise. (Courtesy of the Welch family.)

Wesley Welch reflects:

> *Succession was a progression. Mom and Dad both did a great job of giving us the tools to succeed, step by step. Much credit to them for empowering us. It's never a finish line that you cross; rather, it is a revolving race that you run and a journey that you are on, and you try to keep it going through the next steps to keep your family and business healthy.*

CODA

To conclude, below are some questions to promote reflection, inclusion, and aligned action as everyone moves into the next steps of succession planning together.

For Individuals

1. What role do I realistically want to play in the future of the ranch, and am I prepared for the responsibilities that come with it?
2. What skills or experiences do I still need to develop to contribute meaningfully to the transition process?
3. How can I support the family's goals while also honoring my personal aspirations and boundaries?

For Couples

1. How aligned are we on the vision for our role in the ranch's future, both as partners and as a unit within the family system?
2. What emotional or logistical challenges might arise for us during this transition, and how will we address them together?
3. How can we model strong communication and shared decision-making for the next generation?

For Families

1. Have we defined what "success" looks like for our family and the ranch across generations?
2. What shared values do we want to preserve, and how can we reflect those in our road map and agreements?
3. How will we ensure that every voice is heard, including more reticent or non-operational members?

For Cousins

1. What structures can we put in place now to support healthy collaboration and reduce future conflict between us?
2. How do we ensure fairness and opportunity, even if our roles in the ranch differ widely?
3. What legacy do we, as a cousin group, want to co-create and carry forward?

For Individual Spouses / Partners

1. What concerns or hopes do I have about how the family business affects our relationship and household?
2. How included do I feel in key decisions, and where might I want more clarity or communication?
3. How can I support my spouse / partner while also honoring my own identity, needs, and boundaries?

For In-Laws

1. What is my understanding of the family's goals, and how might I respectfully contribute to or support them?
2. Are there assumptions or misunderstandings I need to clear up about my role in the business or the family?
3. How can I stay involved in a way that promotes unity without overstepping expectations or traditions?

Final thought: Succession is not a checklist; rather, it's a journey. It doesn't happen in a single year or a single meeting. As the Welches remind us, it evolves through baby steps, shared leadership, and trust built over time. Whether you're just getting started or finalizing documents, remember: Every intentional step forward protects the land, honors the legacy, and strengthens the family behind it.

Bibliography

Barry, P., and P. Ellinger. 2010. *Financial Management in Agriculture*, 7th ed. Pearson.

Bentall, David C. 2012. *Leaving a Legacy: Navigating Family Business Succession*. Castle Quay Books.

Bost, J. C., M. J. R. Hoffman, N. Prestopino, and M.K. Lauderdale. 2016. *Estate Planning and Taxation*, 16th ed. Kendall/Hunt Publishing Company.

Dallin, J., J. Hadfield, M. D. Garcia, and R. Larsen, R. 2021. "Utah State University Extension Farm and Ranch Succession Program." *Outcomes and Impact Quarterly* 1 (1): 3.

Dalton, M. A., and T. Langdon. 2022. *Estate Planning*, 13th ed. Money Education Publishing.

Delia, J. G., B. J. O'Keefe, and D. J. O'Keefe. 1982. "The Constructivist Approach to Communication." In *Human Communication Theory*, edited by F. E. X. Dance, 147–91. Harper and Row.

Feirsen, R., and S. Weitzman. 2021. "Constructive Conflict." *Educational Leadership* 78 (7): 26–31.

Gasson, R., and A. Errington. 1993. *The Farm Family Business*. CAB International.

Gomez-Mejia, L., R. Basco, A. C. Gonzalez, and C. G. Muller. 2020. "Family Business and Local Development in Iberoamerica." *Cross Cultural and Strategic Management* 27 (2): 121–36.

González Rueda, A. J. 2017. "Author Literary Legacy: Definition and Elements." *Culturas* 4 (1): 44–68.

Gough, H. G., and A. B. Heilbrun Jr. 1965. "*Adjective Check List* (ACL) [Database record]." PsycTESTS. https://doi.org/10.1037/t02310-000.

Griffin, E., A. Ledbetter, and G. Sparks. 2019. *A First Look at Communication Theory*, 10th ed. McGraw-Hill Education.

Harris, R. 2022. *The Happiness Trap: How to Stop Struggling and Start Living*. Shambhala Publications.

Hauck, J., J. Suess-Reyes, and F. W. Kellermanns. 2016. "Capturing Family Firm Entrepreneurial Orientation: The Role of Family Influence and Identification." *Journal of Family Business Management* 6 (3): 254–69. https://doi.org/10.1108/JFBM-08-2015-0033.

Headrick, J., and A. K. Young. 2023. "Managing Conflict Expectations." In *Developing Human Potential: Personal Approach to Leadership*, edited by Jason Headrick and Lindsay Hastings, 126–45. University of Nebraska–Lincoln Libraries.

Heck, R. K., S. M. Danes, M. A. Fitzgerald, G. W. Haynes, C. R. Jasper, H. L. Schrank, and M. Winter. 2006. "The Family's Dynamic Role Within Family Business Entrepreneurship." In *Handbook of Research on Family Business*, edited by P. Z. Poutziouris, K. X. Smyrnios, and S. B. Klein, 80–105. Edward Elgar Publishing.

Johnson, David W., and Frank P. Johnson. 2021. *Joining Together: Group Theory and Group Skills*, 12th ed. Pearson.

Johnson, D. W., and T. T. Johnson. 2005. *Teaching Students to be Peacemakers*, 4th ed. Interaction Book Company.

Katz, D. 1965. "Nationalism and Strategies of International Conflict Resolution." In *International Behavior: A Social Psychological Analysis*, edited by H. C. Kelman, 356–90. Holt, Rinehart and Winston.

Kay, R. D., W. M. Edwards, and P. A. Duffy. 2024. *Farm Management*, 10th ed. McGraw-Hill.

Kirner, K. 2015. "The Cultural Heritage of Family Ranches: Form, Function and Meaning of Ranching in the American West." *Landscape Research* 40 (6): 648–61. https://doi.org/10.1080/01426397.2015.1074169.

Lencioni, P. M. 2002. *The Five Dysfunctions of a Team: A Leadership Fable*. John Wiley & Sons. Littlejohn, S. W. 1996. *Theories of Human Communication*, 5th ed. Wadsworth Publishing.

Miller, K. 2015. *Communication Theories: Perspectives, Processes, and Contexts*, 6th ed. McGraw-Hill Education.

Muller, Claudio G. 2024. "Chile: Emiliana Vineyard–Family Businesses in the Wine Sector and Sustainability." In *Attaining the 2030 Sustainable Development Goal of Life on Land*, edited by Naomi Birdthistle, 109–17. Emerald Publishing Limited.

Muller, Claudio G., and Fernando Sandoval-Arzaga. 2021. "Conclusions, Lessons Learned, and New Avenues for Further Research." In *Family Business Heterogeneity in Latin America: A Historical Perspective*, 125–44. Springer International Publishing.

Nordqvist, M., and L. Melin. 2010. "Entrepreneurial Families and Family Firms." *Entrepreneurship and Regional Development* 22 (3–4): 211–39.

Obrenovic, B., D. Jianguo, A. Khudaykulov, and M. A. S. Khan. 2020. "Work-Family Conflict Impact on Psychological Safety and Psychological Well-Being: A Job Performance Model." *Frontiers in Psychology* 11: 475. https://www.frontiersin.org/journals/psychology/articles/10.3389/fpsyg.2020.00475/full.

O'Keefe, B. J. 1988. "The Logic of Message Design: Individual Differences in Reasoning About Communication." *Communication Monographs* 55 (1): 80–103.

Picone, P. M., A. De Massis, Y. Tang, and R. F. Piccolo. 2021. "The Psychological Foundations of Management in Family Firms: Values, Biases, and Heuristics." *Family Business Review* 34 (1): 12–32.

Purc-Stephenson, R., C. Hartman, E. K. Marriott, S. Phillips, and C. Scotton. 2025. "Understanding Farmers' Readiness to Develop a Succession Plan: Barriers, Motivators, and Preliminary Recommendations." *Sustainability* 17 (1): 270. https://doi.org/10.3390/su17010270.

Rocklage, M. D., and R. H. Fazio. 2015. "The Evaluative Lexicon: Adjective Use as a Means of Assessing and Distinguishing Attitude Valence, Extremity, and Emotionality." *Journal of Experimental Social Psychology* 56: 214–27.

Sharma. P. 2004. "An Overview of the Field of Family Business Studies: Current Status and Directions for the Future." *Family Business Review* 17 (1): 1–36A.

Silva, M. N. D., and F. S. D. Anjos. 2022. "Generational Succession in Family Farming in the Extreme South of Brazil." *Revista de Economia e Sociologia Rural* 61: e253400.

Silva, T. C., and F. A. Anjos. 2022. "Sustainable Ranching and Innovation in Family Businesses: A Dual Approach." *Agroecology and Sustainable Food Systems* 46 (4): 312–30.

Texas Civil Practice and Remedies Code Sec. 154.023. https://tcss.legis.texas.gov/resources/CP/htm/CP.154.htm#154.

Thorne, M. C. 2021. "Operational Succession in Family Farms: The Need for Structured Role Transition." *Family Enterprise Review* 34 (1): 89–101.

Tracy, K., S. Jackson, and A. Marin. 1994. "Investigating Pragmatic Bases for Managerial Influence: A Comparison of Message Design Logic Perspectives." *Communication Monographs* 61 (4): 307–29.

United States Department of Agriculture (USDA). 2017. "2017 Census of

Agriculture – United States Summary and State Data (AC-17-A-51)." U.S. Department of Agriculture, National Agricultural Statistics Service. https://www.nass.usda.gov/Publications/AgCensus/2017/index.php.

Ward, J. 1987. *Keeping the Family Business Healthy*. Josey-Bass.

Wittman, D., and J. D. Radakovich. 2009. "Positioning Your Ranch for Business Succession." *Rangelands* 31 (2): 4–7.

Wu, J., M. Orr, K. Aizawa, and Y. Inoue. 2021. "Language Relativity in Legacy Literature: A Systematic Review in Multiple Languages." *Sustainability* 13 (20): 11333.

Contributors

Fanni Liu Coward began her impactful career at Texas Tech University in 2009, joining the Department of Curriculum and Instruction before transitioning to the newly formed Department of Teacher Education. There, she focused on teacher education, school partnerships, and community engagement. In 2013, she became a clinical Site Coordinator, bridging university teacher preparation with school districts and mentoring pre-service teachers. She earned tenure and promotion to associate professor in 2015. From 2016 to 2019, Dr. Coward took on leadership roles, serving as Associate Chair and then Chair of the Department of Curriculum and Instruction. A key achievement during this time was integrating community-engaged scholarship into the curriculum. She returned to a full-time faculty role in 2019 and currently serves as the Chair for the Department of Teacher Education.

Jason Headrick's research centers on the crucial intersections of leadership education, rural community development, and inclusive leadership, emphasizing the celebration of diverse identities and contributions. His dissertation notably explored the pedagogy of Massive Open Online Courses (MOOCs) in leadership. An assistant professor of Leadership and Community Development, Agricultural Education and Communications in the Davis College of Agricultural Sciences and Natural Resources, he teaches a range of courses at Texas Tech University, including introductory leadership theory, Effective Groups and Teams, International Leadership, and Leadership and Diversity. Dr. Headrick has actively engaged with university initiatives such as the Institute for Inclusive Excellence and the President's Leadership Institute, and he received the New Faculty Award from the Texas Tech Alumni Association. In 2023, he co-edited and released *Developing Human Potential: A Personal Approach to Leadership*, an open educational resource (OER) textbook, making leadership education freely accessible. He also directs the Civic

Leadership Academy, a program focused on civic leadership development for Davis College students. Dr. Headrick holds leadership roles in various professional organizations, including the Association of Leadership Educators. A native of Kentucky, his background includes deep involvement in 4-H and agricultural life. He earned his degrees from the University of Kentucky and the University of Nebraska–Lincoln.

Patrick Hughes serves as Dean of the School of Professional Studies and a professor of Communication Studies at Texas Tech University. He held the position of Vice Provost for University Programs and Student Success from 2011 to 2022. A distinguished educator, Dr. Hughes is a member of the Texas Tech University Teaching Academy and was honored with the President's Excellence in Teaching Award in 2007. His work has attracted more than $2.5 million in funding. He has also taken on various leadership roles, including former department chair and Associate Dean for Academic Affairs. Dr. Hughes's extensive publications cover communication, the development of interdisciplinary educational programs, and strategic enhancements in both undergraduate and graduate education. He earned a bachelor's degree from Augustana College, a master's from Illinois State University, and a PhD from the University of Denver.

Erica Irlbeck is a prominent figure in agricultural communications, serving as a professor of Agricultural Education and Communications and associate dean at Texas Tech University's Davis College of Agricultural Sciences and Natural Resources. She specializes in risk and crisis communications, alongside video production and strategic communications, earning accolades for her teaching. In 2023, she authored *The Crisis Communications Guide for Agriculture, Food, and Natural Resources*, a textbook reflecting her extensive research. Irlbeck's deep roots in agriculture, stemming from her childhood on an Oklahoma farm and involvement in 4-H and FFA, fueled her passion. Her career transitioned from broadcast journalism and public relations to academia after recognizing a critical need for crisis communication expertise in agriculture. Her research, published in numerous journal articles and presentations, focuses on risk communication, agriculture in media, and curriculum development. Dr. Irlbeck remains actively involved in the agricultural community, holding leadership roles in organizations like the Western Region American Association for Agricultural Education, FFA, and the Lubbock County Farm Bureau. She holds degrees from Texas Tech and Oklahoma State and is married with one son.

Kelly Lange is an assistant professor and Managing Director of Farm Operations in the Department of Agricultural and Applied Economics at Texas Tech University. Before joining Texas Tech, she held positions as an assistant professor at Texas State University and an instructor at Virginia Tech. Dr. Lange is an alumna of Texas Tech, holding a PhD in Agricultural and Applied Economics with a minor in Personal Financial Planning. Her teaching expertise includes farm and ranch management and agricultural finance. Her research focuses on crucial areas such as farm and agribusiness management, agricultural finance, and personal finance.

Mitzi Lauderdale is Vice Provost for Academic Innovation and Student Success at Texas Tech University and an associate professor in the School of Personal Financial Planning. A licensed attorney and Certified Financial Planner™, her teaching and research focus on estate and gift tax planning, charitable strategies, and special needs financial planning. She has published articles and book chapters and co-authored an estate planning textbook. Dr. Lauderdale holds degrees from Texas Tech University and a PhD from Kansas State University. With personal ties to family ranching, she brings a valuable practitioner perspective to financial and estate planning in agricultural operations.

Claudio G. Muller is an Associate Professor of Practice at Texas Tech University's Rawls College of Business, where his expertise lies at the crossroads of entrepreneurship, family business, and organizational strategy. Before joining Texas Tech, Dr. Muller held visiting professorships at Kennesaw State University and Stetson University and he was a professor at the University of Chile. Bringing a global and interdisciplinary lens to his teaching, he specializes in Entrepreneurship, Organizational Behavior, and International Management. Dr. Muller's research delves into how family businesses impact entrepreneurial behavior, innovation, and decision-making, exploring areas like generational transitions, family dynamics, and cultural values. His work reveals how these businesses achieve sustainable success by balancing tradition with innovation. His findings have been published in notable journals such as the *Journal of Family Business Strategy*. Dr. Muller holds a PhD in Business Economics from Universidad Autónoma de Madrid, an MBA from Loyola University Maryland, and a bachelor's degree in industrial engineering from Universidad Técnica Federico Santa Maria.

Amita Quenim is a distinguished clinical psychologist and the Founder of icebreaQers, a family-owned organization specializing in Human Crisis Preparation, Response, and Rehabilitation. With more than four decades of experience, she is a highly sought-after expert in counseling, psychotherapy, crisis intervention, and mental strength and conditioning. Her extensive practice covers a wide spectrum of mental health concerns, including anxiety, depression, personality disorders, stress, addictions, and dispute resolution. Currently, Mrs. Quenim heads the Department of Mental Health at icebreaQers, is part of the National Legal Services Authority in Goa, India, and holds various other professional affiliations. Her expertise extends beyond clinical practice; she is also a columnist, author, and keynote speaker, frequently sharing her insights from psychological research on television. Furthermore, Ms. Quenim has served as a Consultant Psychologist for Dhurandhar Weight Management and as an Advisory Board Member for Manipal Health and Education at Bal Bharati Public School and is the Director of Mandovi Hotels Private Limited, Mandovi Restaurant Services, and the Quenim Family Farm, all family-owned enterprises of the Quenim family, and works closely with her sons Akshay Quenim, founder of Hospitality AQ, and Anish Quenim. Her multifaceted career highlights a deep commitment to mental well-being and effective crisis management for families in business.

Anish Quenim is Director of the Alderson and Griffin Center for Family Business and Entrepreneurship and teaches Family Enterprise at Texas Tech University's Rawls College of Business. He continues to pursue a PhD in Media and Communication at Texas Tech, with a focus on communication between families in business. Mr. Quenim is a TEDx speaker, a certified law enforcement hostage and crisis negotiator, and a credentialed family mediator—efforts directed toward the holistic understanding of family systems and their intersections. Mr. Quenim is also a faculty member at the Maharashtra National Law University in India and teaches a class on evidence-based crisis negotiations at a master's program to better assist legal professionals in mediating crises for families in dispute. He was recently awarded the Jerry S. Rawls Distinguished Undergraduate Educator Award for teaching the family enterprise class, one of the highest recognitions at the Rawls College of Business. Beyond his academic associations, he is also connected to his family enterprise as a generational successor and sits on the advisory board.

Gene Valentini is a highly respected figure in the field of Alternative Dispute Resolution (ADR), serving as Director of the Office of Dispute Resolution Department for Lubbock County. In this pivotal role since 1985, he manages an innovative ADR system that has been instrumental in resolving disputes within the courts of Lubbock County. His expertise has earned him the title of Master of Dispute Resolution for the region. Beyond his county leadership, Mr. Valentini shares his extensive knowledge as an adjunct professor for the Texas Tech University School of Law. Here, he co-teaches the Advanced ADR Clinic, guiding future legal professionals in essential mediation skills. Mr. Valentini also plays a crucial part in statewide initiatives, notably overseeing the USDA mediation program for Texas. A mediator since 1981, he is a seasoned trainer, contributing significantly to the development and implementation of conflict resolution services across the state. His work emphasizes practical solutions and accessible justice for a wide range of legal issues, from civil and family disputes to rural and agricultural matters.

Index

Note: Italicized page numbers refer to figures and tables.

About the Editors

Anish Quenim is Director of the Alderson and Griffin Center for Family Business and Entrepreneurship at Texas Tech University. Mr. Quenim is a faculty member in the Rawls College of Business, where he teaches the university's flagship Family Enterprise course, preparing the next generation of family business successors and entrepreneurial leaders.

Claudio G. Muller, PhD, is an Associate Professor of Practice at Texas Tech University's Rawls College of Business. His teaching and research specialize in strategy, entrepreneurship, and generational transitions for sustainable family business success.

www.ingramcontent.com/pod-product-compliance
Ingram Content Group UK Ltd.
Pitfield, Milton Keynes, MK11 3LW, UK
UKHW040232290926
15471UKWH00002B/144

9 781682 833292